geochemistry and the biosphere:

essays by
vladimir i. vernadsky

First English Translation
from the 1967 Russian Edition
of Selected Works

edited by
Frank B. Salisbury

translation by
Olga Barash

introduction by
Alexander Yanshin, V. P. Emeritus,
Russian Academy of Sciences

synergetic press
Santa Fe, New Mexico

ISBN 978–0–907791–36–2

Library of Congress Cataloging-in-Publication Data

Vernadskii, V. I. (Vladimir Ivanovich), 1863–1945
    [Biosfera. English]
    Geochemistry and the biosphere / essays by Vladimir I. Vernadsky;
translation from the 1967 Russian edition by Olga Barash; edited by
Frank B. Salisbury; introduction by Alexander Yanshin.
        p. cm.
    Includes index.
    ISBN–13: 978–0–907791–36–2 (alk. paper)
    ISBN–10: 0–907791–36–0 (alk. paper)
    1. Geochemistry. 2. Biogeochemistry. 3. Biosphere. I. Salisbury, Frank B.
II. IAnshin, Aleksandr Leonidovich, 1911- III. Title.

QE515.V3913 2007
551.9–dc22

2005054713

Photos reproduced with permission of the Commission on Elaboration of
Scientific Heritage of Academician V. I. Vernadsky, Presidium of the Russian
Academy of Sciences.

Cover image of Crab Nebula courtesy of NASA and the Hubble Heritage
Team (STScI/AURA).

Photo on title page of *Essays on Geochemistry*, courtesy of Dusko M. Du Swami.

The Biosphere by V. I. Vernadsky was first published in 1926 in Leningrad,
USSR. Three years later it was published in France (*La Biosphère*, Paris
Alkan, 1929, 232 pp.). *The Biosphere* was published in Russian for the fourth
time as part of the selected works of V. I. Vernadsky, which included his
publication *Essays on Geochemistry,* under the general title, Биосфера
Biosféra; English: *The Biosphere*, edited by A. I. Perelman, (Mysl Publishers,
Moscow, 1967, 373 pp.)

this book is dedicated to

The late Ganna Maleshka and the late Evgenii Shepelev whose lives were devoted to the study of the biosphere and the evolution of the noösphere.

Vladimir I. Vernadsky (1863–1945)

acknowledgments

The publisher wishes to thank Academician Oleg Gazenko, and the late Drs. Evgenii Shepelev and Ganna Maleshka from the Institute of Biomedical Problems in Moscow for their long-standing friendship and collaboration with Synergetic Press. This collaboration began in 1986 with their support of our publication of the first English edition of *The Biosphere*, an abridged translation from the 1929 French edition. These visionaries of biosphere and noösphere, who pioneered closed system ecological research for improving life on Earth and extending it into space, made it possible for us to make this selection of Vernadsky's writings available in English.

I wish to acknowledge the late Academician Alexander Yanshin, founder of the V. I. Vernadsky Foundation, for his thoughtful contribution to this volume, and to thank the Yanshin family for their help in providing rare and beautiful archival images of Vernadsky and his life. Many thanks go to Academician Eric Galimov, Director of the Vernadsky Institute of Geochemistry for his help in locating historical information. I am also grateful to my dear friend, Dr. Leonid Zhurnya, for his tremendous assistance in communicating with the many contributors to this book in Moscow.

Gratitude is owed to Olga Barash, for the hundreds of hours she devoted to translating Vernadsky's work into English. I am deeply indebted to Professor Frank Salisbury for his detailed editing of the text and for the formidable scientific expertise he brought to bear, which so greatly enhances our understanding of this complex work.

This book would not have been possible without the help and support given by the Institute of Ecotechnics, most especially by one of its directors, John Allen, who introduced me to the emerging field of biospherics, and brought to my attention the importance of Vernadsky's work. John envisioned and cofounded the landmark Biosphere 2 experiment which was highly influenced by Vernadsky's thinking and expanded the frontiers of biospherics.

Other thanks are due to so many people, too numerous to mention, who have donated their time, energy, and skills to this book, especially Eleanor Caponigro for design consultation, Robert Hutwohl of Spirit of the Sun Publications, Anne Visscher, Dr. Mark Nelson, and Linda Sperling for devoted editorial and production assistance.

Why is this book so important now? Vernadsky, who founded both biogeochemistry and biospheric science and greatly influenced Russian and European thought, is nevertheless barely known to the English-speaking world. His writings contain vast knowledge of the history of science, significant insights into the role of humans in the biosphere, and many other ideas vital to the future of life on Earth.

This collection includes the first unabridged third edition of *The Biosphere* (which he continued to revise until the end of his life), and, for the first time in English, selected essays from his seminal book *Essays on Geochemistry.* It has been both an honor and a privilege to publish the work of Vladimir Vernadsky, whose vision and originality still inspires.

Deborah Parrish Snyder

August 2006

contents

the biosphere

the biosphere in the cosmos

the domain of life

a few words about the noösphere

*Vernadsky,
Moscow 1884*

foreword

Editing these two examples of Vladimir Ivanovich Vernadsky's writings through three and more readings has been a most interesting and rewarding experience. Vernadsky's writings are such an enjoyable read because there are several kinds of insights that one may gain from these two works.

There are insights into the status of Earth science and biology during the end of the nineteenth century and the first half of the twentieth century, which was Vernadsky's time – and there are several insights into Vernadsky himself.

He displayed the personality of a scientist almost from the beginning of his life. He was a truly great historian of science. He possessed a unique philosophical mind-set. He was an innovator in his field of geology and a formulator of new sciences. He had a captivating view of the future and especially of the role of humans in Earth's geochemistry and biosphere.

During the 1880s, Vernadsky was associating with several great Russian scientists (as Academician Alexander Yanshin describes in his introduction), especially the renowned chemist, Dmitri Mendeleyev (discoverer of the Periodic Table of the Elements) and Vasilii V. Dokuchaev (the founder of soil science). Thus Vernadsky was on the cutting edge of late nineteenth century science. In 1922, after a number of moves and involvement with the revolution (he resigned from his party in 1918, feeling himself "morally incapable of participating in the civil war"), Vernadsky and his family moved via Prague to Paris, where he wrote most of *The Biosphere* (*Biosfera*), which was published in 1926 in Russian and in 1929 in French (*La Biosphère*). Upon arriving in France in 1922, he was asked to lecture (winter of 1922–23) on geochemistry. It is clear from the *Essays on Geochemistry* that much of the actual writing was done in 1933, but he must have begun his notes in 1922; presumably, the *Essays* were developing during this eleven-year period.

The third edition of *Biosfera* (the edition included here in translation) was published in 1967. When did Vernadsky prepare this third edition? Vernadsky begins his essay on the *noösphere*, which was new in the third edition, by noting that he was writing in 1944 during World War II. The *noösphere* could stand alone, and because its sections are numbered starting with #1, Vernadsky might well have expected it to stand alone. It is a logical conclusion to *Biosfera*, however, and also provides a most appropriate ending for this Synergetic Press volume. Because the third edition contains a considerable amount of material not in the 1926 or the 1929 editions, and because Vernadsky was writing the *noösphere* during World War II, we can assume that he did much editing of the 1926/1929 editions until at least 1944 and possibly until shortly before his death in 1945.

There are two aspects to the insights on the status of science during the end of the nineteenth century and the first third of the twentieth century: the facts and numbers themselves, and what Vernadsky thought about them. Let's consider the facts. Science has certainly progressed and changed during the past three quarters of a century. Thus, one impression during my first reading concerned how many things had changed; that is, I noted the "mistakes" and changes in viewpoint – how much Vernadsky and his contemporaries *did not* know. I'm afraid that I tended to be a bit critical of the facts and numbers presented in these two works. On the second reading, however, I was overwhelmed by the vast amount of knowledge of Earth's geochemistry and the biosphere that *was* known in Vernadsky's time; how much they *did* know! And I was especially overwhelmed by *how much Vernadsky knew*. The breadth of his knowledge of the science of his time is truly amazing – how could one person have so much information at his fingertips?

Nevertheless, one should be careful in reading these two volumes. One should not assume that the presented "facts" and especially the numbers would be the same today (e.g., the amount of some given element in the Earth's crust, in the biosphere, or in the hydrosphere or atmosphere). As Vernadsky was well aware and often states, his facts and numbers were estimates of the time and would likely improve and change in response to further research. I recommend that one should be deeply impressed by the sheer quantity of information of this type that was available back then – and with Vernadsky's ability to bring it together in these two volumes. His summary of the fields is truly a *tour de force*. But one should be wary of the details.

Of course, as a plant physiologist with a now out-of-date minor in geochemistry, for the most part, I could only be aware of mistakes relating to my own field. In several cases, I added footnotes pertaining to some topic that interested me (and often led me to research other sources) – or to some of Vernadsky's terms and usages (and/or those of the translator) that I found difficult to understand, requiring much cogitation and digging into dictionaries. In addition to my footnotes (labeled with "Ed."), the Russian editor of the third edition of *The Biosphere* added a number of footnotes, labeled with (*Comment of the editor of the third edition*). Vernadsky also added several footnotes (no label).

To give you a "heads up" before you delve into the two volumes, following are some examples of points that I had to ponder (others are presented as footnotes):

Vernadsky's thoughts about Earth's history might have been radically changed if he had known about the *oxygen revolution* that is now thought to have occurred about two billion years ago in the Precambrian. Actually, he came very close to discovering it himself (in certain iron rocks). Again and again, he states that the Earth has remained essentially the same since its formation. This cannot be the case if oxygen built up to a high level during Earth's history – such a build-up would change many aspects of geochemistry, not to mention the functioning of the biosphere (which surely was responsible for the oxygen build-up; Vernadsky often mentions that nearly all atmospheric oxygen can only be the product of life, of photosynthesis).

Along with other scientists of his time, Vernadsky assumed that, on an area basis, ocean plankton carried on about the same photosynthesis as land plants. Since the area of the ocean is about twice that of the land, they assumed that, in total, ocean plankton photosynthesize about twice as much as do land plants. We now know that most of the oceans are "nutrient deserts," limited in life by the absence of certain mineral nutrients, especially iron. We think that land plants account for about twice as much photosynthesis as do ocean plankton.

There is a minor matter that struck me as a plant physiologist and would-be ecologist: Vernadsky states that the total area of leaves above a given land area is about 100 times that of the area below (but also counting photosynthesizing organisms on the soil). This "leaf-area index" has now been measured many times and proves to vary from about 1 (deserts) to 11 (tropical rain forests), with a few instances going to 38 (boreal conifers). Averages are about 5 to 8 (typical of deciduous forests).

Perhaps the field that Vernadsky least understood was biochemistry. He never mentions enzymes or genes, and even in his time, the importance of enzymes was becoming known (beginning near the end of the nineteenth century). At one point he even seems to take issue with the importance of proteins (all enzymes are proteins), implying that the concept of "living proteins" is nonsense. By now, enzymes and the genes that control their syntheses are thought to be paramount in life function. Vernadsky does mention that proteins are important, but he never mentions enzymes.

My thoughts and annotations are limited by my knowledge, but an earlier edition of *The Biosphere* was annotated extensively by Mark

A. S. McMenamin.[1] I found the annotations in that edition to be most interesting and valuable. Clearly, McMenamin's knowledge in this field goes well beyond my own.

Both Vernadsky and the translator use some words that might not be familiar to those of us who are not geologists. After much study, I added some footnotes to explain some of these terms. One of them really provided a trap for me: Vernadsky repeatedly speaks of processes, organisms, and chemicals occurring in the *stratis-phere*. At first, being a poor speller, I tried to visualize these things taking place high in the *stratosphere*, but that seemed increasingly preposterous. I went to the dictionaries but could never find *stratisphere*. Considering related terms as well as the context in which Vernadsky used the term, it finally became clear that he was talking about the "sphere" made up of Earth's *strata* – the sphere of *sedimentary rocks*. The term is so insidious that I would have changed it if I could have thought of a suitable synonym, but terms such as *sedimentaryrockosphere* just wouldn't do, so I left *stratisphere* as Vernadsky used it. Be aware!

All these problems are completely secondary to Vernadsky's main theme: *The biosphere is a powerful geological force that has transformed this planet and its geochemistry in a most spectacular way.* I think this must have been apparent to many of his contemporaries (e.g., the science of ecology was vibrant by his time), but there is

---

1. V. I. Vernadsky. 1997. *The Biosphere.* Nevraumont/Copernicus/Springer Verlag, 192 pages. This volume was translated by David B. Langmuir, was annotated by Mark A. S. McMenamin, contains a Foreword by Lynn Margulis with a dozen colleagues, and has an Introduction by Jacques Grinevald. The translation was based on both the French and the Russian editions, published in 1926 and 1929.

probably no other writing produced during that time that pulls it all together as well as these two volumes do.

In his introduction, Alexander Yanshin tells how Vernadsky was fascinated by the world around him from a very early age – how he read avidly in several languages everything that he could lay hands on in his father's large library. Thus his mind accumulated a vast amount of knowledge about the science of his time, as well as the centuries preceding.

Over the years he developed many suitable characteristics for a life of science. He was able to pull together and organize an incredible amount of information and then to apply powers of analysis that were truly phenomenal. As we might well expect, these personality talents of a great scientist led to some highly unique views, relating both to science and how it works as well as to the details of geochemistry and the biosphere. All of this becomes clear in these two volumes.

We see these ideas from the standpoint of a Russian scientist who lived both in his native land and abroad during the period of the Russian Revolution and the Stalin era – which he never mentions in these writings! That in itself provides a perceptive insight into the Russian scientific mind: Although Vernadsky was active in the politics of his day, he leaves all that behind in these writings as he concentrates on questions of the biosphere and Earth's geochemistry.

You'll see that Vernadsky had a tremendous drive, not only to understand the natural world, but to know those who preceded him in seeking that understanding. As he cites and describes the work of hundreds of scientists who came before him, we gain a very broad view of how our modern science has developed.

Today, many scholars who write textbooks or review a particular topic confine their interests to work done in the preceding few years, or at most, few decades. In contrast, as a historian of science, Vernadsky's interests stretch back for a few centuries (especially the eighteenth and nineteenth centuries), revealing that the twentieth century (and now the twenty-first) are not alone in producing good and valid facts and ideas. Furthermore, his admiration for those scientists of previous centuries shines through strongly. Many of us might dismiss the scientists of the age of phlogiston or a geocentric universe, or those who talked about biology and the biosphere (not yet even named!) before the importance of the cell was realized, but Vernadsky could appreciate the creativity and power of their minds and focus on the penetrating ideas that others had overlooked, bringing them to our attention and pointing out that they are reflected in our current thoughts about natural history. One notes Vernadsky's love of history again and again in these writings, but especially in the second of his *Essays on Geochemistry,* in which he reviews the development of his science.

Vernadsky was a philosopher of science as well as a scientist who was concerned with how the universe functions. Some of his philosophical ideas struck me as rather strange (especially in *The Biosphere*) until I was gradually able to fit them into an overall context of his approach to science. At one point he rails against the reliance on mere *hypotheses*, insisting that the only suitable method for a scientist is the (Baconian) system of accumulating facts until *empirical generalizations,* as he calls them, become apparent from the data. Such hypotheses as those relating to the origin of life were "philosophical and religious" hangovers, hardly worthy of true science.

A prime example of an empirical generalization is the Periodic Table of the Elements. When enough was known about the valences and atomic masses of enough elements, it became apparent to his professor, Mendeleyev, that they obeyed a "Periodic Law," and this law was so powerful that it predicted the existence of many more elements. These predictions were fulfilled to a great extent in Vernadsky's time and continue right up to our time. It is noteworthy that Vernadsky was well acquainted with developing chemical theory, based on the atomic models that were only being established with the formation of quantum mechanics; indeed, most of that new science was being worked out during the time that Vernadsky was writing *The Biosphere*.

Although he disparages hypotheses at some points, it is clear from other discussions that he agrees with the modern scientific approach of formulating hypotheses that can be tested by suitable observations and/or experiments. Vernadsky's way of developing empirical generalizations is essentially an inductive method, which is at the heart of most of our modern science. (Since it is impossible to observe every instance of some phenomenon, we must sample the phenomenon and then generalize that our sample is typical of the phenomenon in general. This is induction, but Vernadsky sometimes applies the term deduction to this process. Deduction deduces specific ideas from general laws.)

Vernadsky held to a *substantive uniformitarianism*, or at least a Slavic version of it. The uniformitarianism that we often discuss holds that the present is the key to the past – that the processes going on at present are the same processes that went on during geological history, accounting for the Earth as we see it now. But

Vernadsky's substantive uniformitarianism holds that things have always been the same: that the geology of the Earth has always been the same and that life has always existed on Earth – although he was willing to admit that it has changed by evolutionary processes over geologic time. He holds this view of Earth's history to be one of his important empirical generalizations, but it was probably this view that caused him to overlook the evidence for the oxygen revolution.[2]

This view meant that he strongly opposed the suggestions of his countryman, A. I. Oparin, who in the early 1920s was proposing that life originated on an Earth with a reducing atmosphere (an idea that in modified form is now widely accepted); for Vernadsky, life simply *always existed!* In Vernadsky's writings just before his death, he did accept the concept of an origin of life, but that idea does not appear in these two works.

Unfortunately, the late Alexander Yanshin at first oversold me on Vernadsky as an innovator in his field of geology and as a formulator of the "new sciences" of geochemistry and biogeochemistry. Yanshin claimed so much for Vernadsky that it put me slightly on the defensive. In the process, it made me aware of the incredible number of contributions of Vernadsky's contemporaries and predecessors. Vernadsky quotes dozens, probably hundreds of papers on geochemistry and even biogeochemistry: estimates of the quantity of some elements in the Earth's crust, atmosphere, or biosphere, for example. Such studies are pure geochemistry and biogeochemistry. My thought then was, how can we say that

---

2. *Ibid.*, p. 40–41. In one of his annotations, McMenamin describes systematic or substantive uniformitarianism in considerable detail.

Vernadsky founded the sciences of geochemistry and biogeochemistry when this vast amount of work in these fields had
already been done? Clearly, Yanshin gave Vernadsky credit for his
synthesis of the ideas of these fields. This synthesis was indeed a
highly original and significant contribution. (Yanshin mentions
a number of other sciences as examples of Vernadsky's synthesis;
my familiarity with Vernadsky's works – his many, many publications – is not sufficient for me to judge his contributions to those
sciences.)

As Yanshin points out, Vernadsky seldom claimed that he was
the first to develop a field. He gives credit for the concept of the
biosphere to Eduard Suess, who published the term in 1875.[3] My
geochemistry textbook published in 1952 gives Vernadsky credit for
the term *biogeochemistry* but says: "The concept of the biosphere
was introduced by Lamarck [near the end of the eighteenth
century]. . . ."[4] Vernadsky also mentions Lamarck as the originator
of the *concept* but not the term. In short, for some time there have
been those who realized that Earth's organisms and the physical environment with which they are closely associated should
be recognized as a very special part of this planet, but it was
Vernadsky's concept of *living matter,* a term that he uses over and
over, that strongly emphasized the effects of life on the Earth. We

---

3. *Ibid.*, p. 23. McMenamin gives a translation of the quote from Suess, who
speaks of a sphere of organic life with roots in the lithosphere and organs
in the atmosphere "to breathe." The quotation ends: ". . . we can distinguish
a self-maintained biosphere [eine selbständige Biosphäre]." This is Suess's
only use of the term.

4. Brian Mason. 1952. Principles of Geochemistry. John Wiley and Sons, Inc.
New York. See p. 193.

only have to think of Mars and Venus to realize the impact of the biosphere on our orb.

Incidentally, I found it interesting that Vernadsky seldom if ever referred to his own laboratory or field work. He must have done much "hands-on" science, but his approach in these two volumes is that of the observer, the synthesizer, who bases most of his synthesis on the work of others.

In any case, it is clear from Yanshin's introduction, if not from the texts themselves, that Vernadsky's influence was great, particularly in his Russia, even as this influence was slow to penetrate to the Western world.

One of Vernadsky's most fascinating concepts encountered near the ends of both *The Essays on Geochemistry* and *The Biosphere*, is his view of the future and especially of the role of humans in the Earth's geochemistry and biosphere. Humans, with their science and technology, have changed and are continuing to change the nature of the biosphere. The end result will be the *noösphere* (from the Greek: *noó(s)*, mind). As with the term *biosphere*, Vernadsky credits another author, Edouard Le Roy, with the term *noösphere*. Another source says that Vernadsky, Teilhard de Chardin, and Le Roy jointly invented the term in 1924. In any case, the concept of the impact of the human mind on Earth seems obvious to us and has probably been obvious to some extent to many others, especially during these years of industrialization and technology, but Vernadsky says it particularly well.

some notes on editing these manuscripts

In attempting to copy edit these manuscripts, it has not been difficult to correct matters of grammar, punctuation, and style. This
was done according to established rules outlined in various style
manuals. But because the manuscript with which I was asked to
work was a translation, and I did not have access to the Russian
original, many special problems arose. Often a word or a sentence
did not sound correct to the ear of a native English speaker.
Sometimes in such cases of confusion, I left the translator's word
because it sounded a bit more Russian but was still appropriate
enough for easy understanding. After all, Vernadsky wrote in
Russian![5]

In some cases, I referred to the 1926/1929 version of *The Biosphere*
translated by Langmuir, who was a native English speaker who also
had an excellent command of Russian, French, and science. His
choice of words was often very helpful to me. Mark McMenamin's
annotations in the Margulis edition were often helpful because
they explained Vernadsky's ideas or clarified various scientific ideas
– sometimes confirming and sometimes rejecting my suspicions
about changes that should be made in Barash's translation. However,
the 1967 Russian edition (from which the present version was translated) had many additions to the 1926 version, and sometimes my
problems were in a portion of the text that did not appear in the
1926 version. Also, I had no access to another English version of *The
Essays on Geochemistry* save for Barash's translation.

---

5. There may be better alternative English words to render Vernadsky's ideas
but we have done our best to work with what the intended meaning was
from this translation.

Langmuir states that his revision of the Russian/French version was "a rather drastic one." That is, he probably took more liberties with the language than did Barash. Hence, this version, in spite of my editing, is probably a more literal translation than is the Margulis edition – which nevertheless accurately preserved Vernadsky's meaning, according to Langmuir.

A good example of the problems of editing a translation by a non-native English speaker is the use of the articles "a," "an," and "the." These articles do not exist in Russian, so a Russian translator translating into English must insert them where needed – a very difficult task. Hence, I considered them "fair game" – to add or delete as seemed appropriate to my ear. The rules for placement of the articles in English are tricky to say the least. Although such rules do exist, they are not easy to apply. Yet a native speaker whose ear is attuned to the language seldom uses them improperly. In my editing of this translation, I must have removed a few hundred definite articles ("the"), mostly before plural words (e.g., *solar rays*) or compound nouns (e.g., *green living matter*). I also had to add quite a few definite and indefinite articles when it seemed appropriate. I apologize if these additions or deletions were not always what they should have been.

Having just read the page proofs for the fourth time, I continue to be deeply impressed with the two volumes, especially for two reasons. I am still amazed by what was known by the first third of the twentieth century. Most of the important geochemical and ecological themes being discussed now are foreshadowed in these volumes. One example: I thought that perhaps Vernadsky's time had not been aware of the potential problems, particularly global warming, posed by the rapid increase in carbon dioxide thanks to burning of fossil fuels. But there it is on pages 189-190 where Vernadsky considers the views of Arrhenius.

The breadth of Vernadsky's knowledge of the literature of geology, geochemistry, the biosphere, the kinds of living organisms, and many other topics is simply mind boggling. I began to wonder: *Could anyone alive today duplicate Vernadsky's feat, but this time incorporating all the information that has been added to and expanded upon what was known in his time?*

This work has been enjoyable as well as challenging, and it was made possible by Deborah Parrish Snyder of Synergetic Press. She has been a wonderful help and a joy to work with.

Frank B. Salisbury
Professor Emeritus of Plant Physiology
Utah State University, Salt Lake City

Vernadsky,
Moscow 1940

introduction

The purpose of this introduction is to familiarize the reader with the powerful historic figure of Vladimir Ivanovich Vernadsky (1863–1945) as a great scientist and thinker of the twentieth century. The scope of his genius can be fully comprehended only through acquaintance with all his creative work in the fields of natural science, biology, and philosophy, which by far exceeds the common idea of Vernadsky as a geochemist, mineralogist, and geologist.

Vernadsky's teachings on the biosphere and noösphere belong to science, just as Darwin's theory of the evolution of species, Bohr's fundamentals of quantum physics, and Einstein's relativity theory. That is why this edition is an homage to the history of fundamental scientific ideas to which the teaching of the biosphere clearly belongs. Vernadsky was the founder of genetic mineralogy, geochemistry, biogeochemistry (the concept of "living matter" as a geological force), the theory of the biosphere, radiogeology, and hydrogeology. His ideas gave birth to many scientific disciplines. By force of logic and generalization he anticipated the ideas of unity of time and space, of the physical vacuum and of the asymmetry of space. His ideas of the "local" features of sections of the world's ocean, occupied by living organisms or growing crystals, have not yet been fully understood and developed in terms of present-day physics. Long before World War II, Vernadsky had written about the potential use of atomic energy for military purposes and, in this connection, about the great responsibility of scientists, though physicists had not even thought about creating an atom bomb. Such was the scope of his thought and vision.

Vernadsky's teachings not only prepare the ground for planetary thinking, but also exemplify a full-scale understanding of the unity of the planet's living and non-living nature and the unity of the planet with its cosmic environment. This unity is the gist of Vernadsky's teachings.

V. I. Vernadsky is undoubtedly a great and rare phenomenon in the history of natural science. Such powerful figures do not emerge every century. This is the way I see him, and this is the way I would like to introduce him to the English-speaking reader.

vernadsky's life

Vernadsky, St. Petersburg, 1875

The future scientist and Academician Vladimir Ivanovich Vernadsky was born in St. Petersburg into a nobleman family with ancient historic roots in the Ukraine. In his early years, he was an ordinary boy, a bit phlegmatic and shy, and manifested no signs of genius. From early childhood, he was keen on reading. No one in the family controlled his reading, and he used his father's large library to his heart's content. At age twenty-three, he recollected in one of his letters:

I threw myself at books early and read voraciously everything I came across, constantly digging in my father's library. . . . From these early years, I especially remember various books on geography, not only

about travels but also rather dry books that seemed difficult for my
age, for instance, The Earth by E. Reclus. . . . At the same time, I was
fond of books on history, especially Greek.

And then, speaking about his High School years:

I was deeply interested in the history of the Church. . . . My home life
gave me the main thing: dozens of journals, Russian and foreign,
that my father subscribed to.

Vernadsky's father was a professor of political economy, which
seems very far from geological sciences. But political economy
compares human needs with natural conditions; from here, it is not
far to Vernadsky's subsequent understanding of nature and man's
place in it. As a young man, Vernadsky wanted to take up history but
decided first to get an education in natural science. In 1885, Vladimir
Ivanovich graduated from the natural science department of the
faculty of Mathematics and Physics of St. Petersburg University, and
continued at the faculty to prepare for a professor's degree.

Vernadsky studied at St. Petersburg University when it was
in its heyday: a brilliant constellation of scientists gathered
there; they created an era not only in Russian but also in world
science. His teachers were the chemists Mendeleyev, Butlerov,
and Menshutkin, the soil scientists Dokuchaev and Kostychev,
the geologist Inostrantsev, the geographer and meteorologist
Voyeikov, and other famous scientists of that time. Each of them
made a great contribution to twentieth-century science. The first
among them was D. I. Mendeleyev. I do not need to introduce the
creator of the Periodic Law and the Periodic Table of the Elements,
which are studied in every school. The ideas of Mendeleyev, and

especially those of the soil scientist Dokuchaev, greatly influenced Vernadsky's later scientific work.

Having received a geological education, Vernadsky first took up crystallography and mineralogy at St. Petersburg University. After moving to Moscow, he delivered lectures in mineralogy at Moscow University, at the chair of a famous geologist and subsequent Academician of his time, A. P. Pavlov, who was one of Vernadsky's teachers whose name we shall come across below.

During his student years and his work at Moscow University, Vernadsky took part in Dokuchaev's expeditions, studying soil

Vernadsky and other students at St. Petersburg University 1884

chemistry in different regions of Russia. It is easy to understand that the science created by Vernadsky – geochemistry – turned out as "genetic" as Dokuchaev's soil science. It embraced not only the distribution and content of chemical elements in the Earth's crust, the atmosphere, and the natural waters, but also their origin under different conditions and the places of their existence, their migration in the course of geological processes, and especially their biogenic migration as the result of the activity of living matter in the biosphere. That is why the titles of separate sections of *Essays on Geochemistry* contain the word *history*: history of carbon, of oxygen, and so on.

Although the scope of his scientific work was tremendous, Vernadsky never limited himself to it. Like many representatives of the Russian intelligentsia of his time, he was deeply concerned with social and political problems. He plunged into social activities early, in his student years. He was one of the founders of the first political party in Tsarist Russia – the Constitutional Democrats – and a member of its leading central committee. Twice he was elected a member of the State Council, the supreme elected body of Russia, where he expressed his emphatically democratic political views. In 1911, he resigned from Moscow University, along with twenty-one leading professors, in a collective protest against the Education Minister's arbitrary rule. He then decided to give up teaching and to devote himself entirely to scientific work. After 1917, he gave up political and social activities as well.

All of Vernadsky's scientific work was accompanied by extensive organizational activities: He attracted the interest of the Academy of Sciences, with its potential for scientific investiga-

Vernadsky and other profes- sors of Moscow University who resigned in 1911 in support of students' protest against the Education Ministry

tion, to the circle of scientific problems he was anticipating, or he created new branches in the Academy. In 1912, he founded the first radiochemical laboratory in Russia. In 1915, on his initia- tive, a committee of the Academy of Sciences was created to "study the natural productive forces of the country." At first, it was meant to discover new sources of strategic ores, because Russia was taking part in World War I. He also included the study of uranium ore deposits as a task of the committee. He was chairman of this committee for fifteen years, until it became the State Geology Committee.

In 1926, at Vernadsky's suggestion, the "Committee on the History of Science" was founded at the Academy of Sciences; Vernadsky remained its head until 1930. It later became the Institute of History of Natural Science and Technology which continues to carry out successful work together with a similar branch of the Smithsonian Institution in the United States.

During the Russian Civil War 1918–1921, he actively participated in the creation of the Ukrainian Academy of Sciences and became its first president. No matter where he lived during the most difficult years, he created new branches of scientific research, groups of scientists, and laboratories that proved long-lived because he founded them on new, fundamental scientific concepts and perspectives.

vernadsky's teachings

Vernadsky's idea that living beings possess a great geological significance in changing the Earth's face gave birth to a new science – biogeochemistry. This concept increasingly interested Vernadsky and eventually became the main content of his creative work. Near the end of his life in 1944, he wrote as if summing up:

I spent the years of World War I in constant research and creative work, and I have been going on in the same direction up till now. . . . All these years, no matter where I was, I was captured by the thought of geochemical and biogeochemical manifestations in the surrounding nature (the biosphere).

These new geochemical and biogeochemical ideas did not enter the scientific mind of that time. Old notions reigned supreme in geology many years after Vernadsky's works had been published. His thought was far ahead of his time, and he was not understood by many of his contemporaries. That is why the creation of biogeo-chemistry and the concepts of the biosphere not only manifests his scientific genius, but is also a striking example of the anticipa-tory power of this scientist, of his persistence in reaching the goal, of faith in his ideas, and of being able to work in most unfavorable conditions. We should remember that his most significant scien-tific achievements were made during the years of civil war and economic breakdown in Russia.

The central concept of Vernadsky's teaching is that of the biosphere, but its definition in literature has been vague until now. Many people define it in an easy way as "the realm of life," the territory of the planet inhabited by living organisms at any given time. But it is not quite like that. In Vernadsky's under-standing, the biosphere is a historic concept. It dates back to the very first manifestations of life on Earth – manifestations that

Vernadsky in suburbs of Prague (circa 1928)

created the oxygenated atmosphere and changed the planet's surface in the course of life's evolution, which is still ongoing. By "the biosphere," Vernadsky meant all the layers of the planet, and first of all the layers of the Earth's crust, that had undergone the influence of biogeochemical activity throughout its entire geological history. This idea of the historic character of the biosphere was shown rather recently in a large geological, geochemical, and paleontological work, a book by the Leningrad geologist Andrey Lapo that was translated into English as *Traces of Bygone Biospheres* (Synergetic Press, 1987).

Vernadsky's concept of the biosphere is as diverse and hierarchical as the structure of the actual biosphere. This concept integrates the data of all sciences that relate to the Earth, all biology, chemistry, and biochemistry. The growing specialization of the sciences has provided great progress in the knowledge of profound details and intimate mechanisms of life, without getting closer to understanding its essence. This widely acknowledged drawback of contemporary science consists in its failure to embrace Earth's nature in all its scope and profundity, on a planetary and cosmic scale. Given this background, the concept of the biosphere, in its unprecedented scope and depth, is an excellent example of the integration of the sciences and a great event in the science of the twentieth century.

All this became possible due to the long-term and conscious approach to the integral study of any phenomenon in connection with other phenomena. Vernadsky wrote about the necessity of such an approach as long ago as the end of the nineteenth century, in his student diary; at that time the process of the specialization of science and its division into a number of specific disciplines and trends was only beginning. Later, in 1920, he privately recollected:

I have long been surprised at the lack of desire to embrace Nature as a whole in the field of empirical knowledge, whereas it is within our grasp to do so. Often we as scientists give only a mere collection of facts and observations where actually we could present the whole . . . It looks like some mental laziness. We feel that if we make an effort, we can rise to embracing the phenomenon as a whole, but this effort is not made, and judging by the literature nobody makes it.

Only much later, in the second half of the twentieth century, was this approach in science widely realized, acknowledged, defined and called integrative, systematic, global, etc. An example is provided by one of the fundamental principles of biology; namely, the unity of the organism and its environment. The briefest definition of this principle was probably given by the Russian physiologist I. M. Sechenov, who believed in the late nineteenth century that the description of an organism would not be complete without the description of its environment. Vernadsky came to affirm this principle on a different scale, independently and in his own way as a geologist and a biochemist, through ideas about the matter-energy connections of organisms and their environment in the biosphere. But he did not restrict the concept to an individual; he thought on the geological scale of the living matter of humankind as a whole. He wrote:

Humankind, as living matter, is constantly connected with the matter-energy processes of the defined geological envelope of the Earth – its biosphere. Not for a single moment can humankind be physically independent of its environment.

Vernadsky in his St. Petersburg office 1921

In the course of the development of Vernadsky's notions of the biosphere as the most active part of the planet's matter and its connection with cosmic factors, he was getting a more extensive idea about its limits – the boundaries of the integral approach within which the biosphere processes were to be studied in order to get as close as possible to an exhaustive understanding of the subject. Reading his *Essays on Geochemistry* and *The Biosphere*, it is easy to see that the author, while projecting the phenomena he considers on the whole biosphere, is constantly reminding us that the planet itself also has its external connections, its "habitat" in the cosmos. The dependence of the biosphere upon the luminous radiation of the Sun is obvious to us all, but Vernadsky had left the problem of the Earth's cosmic connections open for further discoveries in this field, since he understood their scientific inevitability. We now know more about these connections than in Vernadsky's time. It is enough to remember the galactic and extra-galactic cosmic radiation, the role of the magnetosphere of the Earth in preventing the destructive effects of cosmic radiation upon terrestrial life, the loss of terrestrial oxygen in the upper layers of the atmosphere, the role of the ozone screen of the Earth, and the connection of biological processes with solar activity. But all this is included in the realm embraced by Vernadsky's scientific mind, which actually had no boundaries.

Reflecting on the structure or the macrostructure of the visible cosmos as an object of scientific study, Vernadsky clearly distinguished "three separate layers of reality," within which the scientifically stated facts are situated. These three layers of reality, in all probability, differ distinctly from each other in properties of space and time. They penetrate into one another, but they are definitely realms unto themselves, distinctly delimited from one another both in their content and in the methods of studying their manifestations. These layers are the following: the phenomena of cosmic spaces, the planetary phenomena of our visible "nature," so close to us, and the microscopic realm in which gravity is of secondary importance. The phenomena of life are observed only in the two latter layers of world reality.

Vernadsky embraced with his mind's view all these layers of the world's reality. Such scope of thinking is unprecedented in the last centuries of exact (not speculative) sciences. At the same time, one cannot but think that the scope of his ideas is directly related to philosophical and religious systems of the ancient East, India, and Tibet, although Vernadsky based his ideas strictly on empirical data from which he never digressed. In the literature "the phenomenon of Vernadsky" is often spoken of in connection with the power of his scientific thinking. Of course, there were many factors that contributed to his development as a scientist, such as his education and upbringing, his will, his character, his persistence (as noted, he read in fifteen languages), his analytical and critical mind, etc. But many other scientists possessed all these to a certain extent. What, then, is the "phenomenon of Vernadsky?"

The main features of his thinking are probably his wide range of thought and his extraordinary breadth of scientific generalization, not in some definite field but in everything with which he dealt. The scope of his scientific generalization is enormous. In order to embrace with his mind the specific role of living organisms in changing the Earth's crust, which he had already comprehended in general, he had to abstract himself from the specific functions of millions of animal species, and see only the one common function involved in changing the surrounding non-living nature – a function that could not be seen by botanists, zoologists, nor microbiologists. He saw this in the geologically generalized property of all living organisms – in their capacity to transform the energy and matter of the upper envelopes of the planet. It was then that the new, large-scale notion of "living matter" appeared to denote this generalized bearer of the generalized function of living organisms.

Being a geologist, Vernadsky could not do without this notion of his newly realized geological power of living organisms, comparing it with the classical tectonic processes: volcanism, water and wind erosion, and other traditionally known geological causes of change of the planet's face in its geological history. This newly discovered geological force has turned out to be much more powerful than the natural elemental forces, and it is of cosmic essence, since it is brought into action by the energy of cosmic and solar radiation.

These new notions and terms in science each have a different degree of significance. The notion of living matter has invaded  the very structure of natural sciences. It has made the vast world of living organisms, the world of botanists, zoologists, and microbiologists, an object of a quite different geological science. Geology has drawn from biology a new geological factor that had never been included into its competence, and which is, as it has turned out, the crucial one among the other, traditional, geological factors changing the planet's face. Now, botanists and zoologists as well as microbiologists are able to consider their objects from a different standpoint, that of their geochemical function in the biosphere.

Vernadsky circa 1911

Vernadsky's work on the role of living matter in Earth's history did not stop at this point. With his mind's eye he managed to see the part of living matter that, though relatively small, was extremely important from the standpoint of geology: the living matter of humans. Before civilizations appeared, the human population of the Earth had not differed from the whole mass of higher organisms in its biogeochemical role. But it began to manifest itself as an essentially new natural force in Mesopotamia, Chaldea, and Athens, in the school of Plato, in the philosophy of Democritus, Aristotle, and Socrates, in the teachings of Ptolemy, and then, with an increasing crescendo, in the teachings of Copernicus, Bruno, Newton, Darwin, and Einstein. I should add Vernadsky to the list, too. Of course this list is quite relative, and intended only to illustrate Vernadsky's thought that such a small part of the biosphere generated and is still generating a qualitatively new factor in the development of the biosphere: the rapidly increasing sum of scientific knowledge about the biosphere, about the direction and volume of humans' productive activity, which had reached the scale of a new geological force.

This empirical fact brought forth another generalization unprecedented in the history of science, but characteristic of the scope of Vernadsky's mind. As in the case of the global notion of "living matter," he abstracted the essence from the specific content of the countless specific scientific facts in the numerous particular scientific disciplines. In this vast multitude of scientific data he saw a certain general essence and called it "scientific thought." All the diversity of science of all times and nations was generalized into "scientific thought," like all the diversity of life into "living matter." By this name he denoted this qualitatively new and, again, geologically significant product of biosphere development as a generalized and independent force on a geological scale, this time produced by an extremely small quantity of the planet's living matter – that of humans.

In his notion of "scientific thought" or "scientific mind" he saw not an encyclopedia of science, but a generalized, average

motion of human thought as part of the planet's development
– its geological history. That is why he collected his unpub-
lished notes under the title "Scientific Thought as a Planetary
Phenomenon," which he failed to complete although he
considered it his principal book. These notes were later
published under the same title (Mysl, 1991).

Taking into consideration the rapid increase of human
activities affecting the biosphere, and the anticipation of
their further increase, Vernadsky came to a conclusion about
the appearance of a new qualitative state of the biosphere, in
which "scientific thought" increasingly becomes the main factor
determining its further state and evolution, and is already an
independent factor of the biosphere determining and directing the
practical activities of humanity in nature and society. He named this
new stage of the biosphere's development the "noösphere," which
means the sphere of science-based intellect, of a new attitude of
humanity towards its environment. He believed that the humans
of his time had already entered or were entering this new state of
the biosphere.

It has been noticed that Vernadsky avoided introducing new
terms into science. If necessary, he found them in the scientific literature, which he knew very well. This happened also
with the term "noösphere," which was suggested by the French
mathematician and philosopher Le Roy in 1927. Vernadsky used
Le Roy's term in his paper, although he attached to it a more
comprehensive meaning.

The creation of the teaching of the biosphere coincided with
the situation of that time in traditional geology, when the

increasing influence of Man upon Nature had reached the level of a geologic force. That is why the geologists had to find a proper designation for the contemporary stage of the planet's development in terms of conventional geochronology. Since our school years, we have known that the development of life on Earth passed through long geological eras quite different in content: the Paleozoic (and now, as we have come to know, earlier eras as well), Mesozoic, and Cenozoic eras. The latter formally is expanded to our time, but it also embraces the last millions of years in the Earth's development when some ancestors of contemporary Man appeared, separate centers of primeval human society were springing up slowly but inevitably, and the first centers of civilization arose, later blending into the all-human civilization that has embraced the planet from pole to pole.

To designate the contemporary stage of the Earth's geological history, the term "anthropogenic era" was suggested in 1922 by the geologist A. P. Pavlov, one of Vernadsky's teachers. Another term for the same purpose, Psychozoic Era, was suggested by the American geologist Charles Schuchert. In both variants the main factor, the backbone of the contemporary geological epoch, was Man. No doubt, the present and future history of the biosphere's evolution will be written by humankind, though this part of the planet's "living matter" (biomass) is insignificant in its percentage of the total biomass. I cannot say how it will act on our long-suffering biosphere. Maybe ichthyosauruses will not be the last once powerful but now extinct species of Earth's inhabitants.

Vernadsky could have meant this when he wrote as early as 1902 about the great responsibility of scientists for their activities: "At present in the field of exact knowledge, we are standing on the border, on the verge of great discoveries. . . . Cannot the forces discovered by nature be used to do evil and harm?" Later, in 1922, he put it more definitely:

We are approaching a great revolution in humanity's life, which cannot be compared to anything in the past. The time is coming when Man will be able to control atom energy, a source of power that will give him an opportunity to build his life as he pleases. . . . Will he be able to use this power, to direct it to good, not to self-destruction? Is he mature enough to manage this power which is inevitably to be given to him by science?

The reader will understand the power of Vernadsky's anticipation, taking into account that these words were written when physi-

Vernadsky on vacation in Peterhof, 1931

cists, including Niels Bohr's "brain center," did not even think about the actual use of atomic energy.

But let us return to geochronology, with which we were discussing the name of our current geological era. In the literature, one can come across attempts to compare or even oppose the names "anthropogenic" and "psychozoic" era to the notion of the noösphere. This is an obvious misunderstanding. These notions are incomparable in principle, since they lie in different planes of thinking. They belong to different "fields of reality," in Vernadsky's terms. The notion of the noösphere, as well as that of the biosphere, lies outside geochronology. Vernadsky's "biosphere" embraces all the geological eras related to the activities of living matter since Precambrian time, to which the first manifestations of the activity of micro-organisms in the ancient ocean date back. This means that this notion embraces all geological eras and cannot refer only to their last stages. The notion of the noösphere has nothing to do with geochronology. It means only the new, contemporary stage of development of the biosphere. At this stage, the global cycle of the planet's matter and the transformation of solar energy in the Earth's envelopes become essentially dependent on the increasing sum of knowledge – the scientific and subsequent practical activities of mankind. Thus the noösphere should be understood not as something new, but as the present, current state of the terrestrial biosphere in the contemporary geological era, no matter whether it is called "psychozoic" according to Schuchert, or "anthropogenic" according to Pavlov.

vernadsky as historian

The history of natural science, and of science in general as the history of the human scientific mind, was the second great scientific interest of Vernadsky. The work done by him in the field of natural sciences seems to be sufficient for several scholars. But he also was one of the twentieth century's most outstanding historians of science. Working at his specific problems of crystallography, mineralogy, and geology, he constantly went beyond the limits of the studied area to the vast spaces of the greater history of knowledge, and not accidentally but quite consciously:

More and more am I carried away by the idea of devoting myself seriously to the history of science. But it is hardly possible: I feel a lack of education and an insufficient power of mind for such a task. Such work will take up many years, as I shall have to prepare for it for a long time. (From a letter to his wife N. V. Vernadskaya, 1893).

But this was not mere intention. In 1902–1903 he delivered lectures on the history of natural science at Moscow University which later were published as a separate book based on the archive records. Here he reveals himself not only as a professional investigator of the history of specific areas of the natural sciences, but also as an outstanding theoretician in a field of historic knowledge that was still nascent.

The very first pages of *Essays on Geochemistry* show Vernadsky as an historian. His works in general are noted for the scrupulous search for and descriptions of the historic predecessors of his own or other ideas, which, unfortunately, is not characteristic of most scientists of the present-day generation (though it is always characteristic of real scientists).

I shall give one of his notes to *Essays on Geochemistry* as an example of his attention to the history of scientific thought and to particular scientists. This note is devoted to a Croatian scientist of the eighteenth century, R. J. Boscovich:

R. Boscovich – Jesuit and citizen of the Dubrovnik Republic. Not an Italian, as it is sometimes stated, and against which he had always protested. . . . He became a French citizen in 1773, and passed several years (1773–1782) in Paris as an academician and Director of "Ortique de la Marine." Being a Jesuit, he had numerous friends and influential enemies (including d'Alembert). It is interesting to note the sharply opposite estimations of his scientific importance in the nineteenth and twentieth centuries. For us he is one of the greatest scientists; but in 1841 an outstanding astronomer, F. Arago, considered him "a person to whom Lagrange and d'Alembert related with great contempt" and "a mediocre foreigner" (F. Arago, Oeuvres completes, ii, p. 139–140, 1854). The interest in Boscovich has begun to increase since the middle of the nineteenth century in connection with the rise of the new physics; but it had never flagged in the previous century either.

This small note reveals Vernadsky's typical approach to historic material, the scrupulous manner of looking into the personality of a scientist, the conditions of his work and his scientific and social surroundings. Of course it was promoted by the fact that he read in fifteen modern languages – all the Slavonic, Roman and German languages. No doubt he knew the classical ones, too.

thinker

Vernadsky was well-read in the issues of philosophy. He studied and knew not only the philosophical systems of the West, but also those of ancient India and China. He did not think it possible for himself to prefer any philosophical system he knew, and thus he remained a skeptic. He contended that the foremost task of a scientist, studying reality, is accomplished not by philosophy but by empirical science, by the specific facts it uncovers, and by empirical generalizations based on these facts.

*Vernadsky,
Moscow 1940*

He considered the notion of an "empirical generalization" to belong to the highest category of scientific cognition, unlike scientific hypotheses and theories, which always turn out to be temporary. As a specific example of an empirical generalization of everlasting significance, he gave the Periodic Law of Chemical Elements discovered by Mendeleyev. To this we can now add his own empirical generalizations, such as the ideas of "living matter" and "scientific thought," which need no proof. His teachings on the biosphere and the noösphere may probably also may be counted as such.

Vernadsky wrote a lot about the significance of science as a whole; scientific hypotheses and theories on the one hand, and of empirical generalizations on the other hand. He noted the lack of philosophical understanding concerning the issue of the place of empirical generalizations in contemporary cognition theory, and was the first to put forward this problem, which has not attracted the attention of philosophers up till now.

The volume of Vernadsky's manuscripts on philosophical issues is large enough, but these papers were never published during the author's life. He did not intend them for publication, since, it seems to me, he perfectly understood their discrepancy with the officially cultivated dialectic materialism, which had acquired the status of State Philosophy.

Vernadsky had always stressed the distinct difference between the cognitive potential of philosophy and that of the empirical sciences. But even in the very first manifestations of the ancient philosophical mind, he saw the dawn of scientific knowledge. He wrote in one of his letters of 1902:

I look at the meaning of philosophy in the development of knowledge in quite a different way than most naturalists, and attach to it great fruitful significance. I think these are aspects of one and the same process, aspects that are absolutely indispensable and inseparable. They can be separated only in our minds. If one of them decays, the other will stop growing too."

And here are words from his 1902 lecture, "On a Scientific Worldview":

Never in history have we observed science without philosophy, and studying the history of scientific thought, we see that throughout all the time of its existence, philosophical concepts and philosophical ideas have permeated science as an indispensable element.

This is an example of his own philosophical reflections from his student's diary of 1885:

What is time and space? These are problems that throughout the ages have interested the human mind in the form of its most brilliant representatives. . . . No doubt, time and space do not exist in nature separately; they are inseparable. We know not a single phenomenon that covers no time and no space. Only for logical convenience do we imagine time and space separately, only because our mind is used to doing so while solving problems.

In reality we see time and space separately only in our imagination. To what do these inseparable parts belong? They belong to the only thing existing, to matter, which we divide into two basic coordinates: time and space.

These words were written more than a quarter of a century before Professor Minkovsky, at the Mathematical Congress in

Vernadsky in his office, Moscow 1940

Cologne in 1908, overwhelmed listeners by his new ideas about the single, indivisible concept of space-time, about time as the fourth dimension of space. Such was the force of Vernadsky's anticipating thought in a field that was not even his specialty.

He always tried to understand Newton's abstract geometrical and mathematical space through real parts of physical space covered by natural bodies, including living organisms that not only "cover the space" but also form its characteristics in the space they cover. This is a way of understanding Vernadsky's thoughts about the difference between the space of natural sciences and the space of philosophy and mathematics. A large section of his "Philosophical Thoughts of a Naturalist," entitled "Time and Space in Living and Inert Matter," is devoted to this problem.

His ideas of the space of living organisms were drawn from L. Pasteur's discovery of spatial difference and dissymmetry in organic molecules – which, as it turned out later, manifests itself also in macrophenomena, such as the direction of convolutions of spiral shells of mollusks, the spirality of some tree trunks and climbing plants, right- and left-handedness of man, different functions of the right and left hemispheres of the brain, and so forth.

From separate empirical facts, Vernadsky draws the conclusion about a specific state of the real physical space of living organisms. He sees this specific state in the asymmetry of processes and structures on the territory occupied and controlled by organisms. Thus he posed the philosophical problem of the real characteristics of space (unlike the abstract space of geometry), separate parts of which are controlled by living organisms with their selective asymmetrical behavior, unlike the non-living nature around them in which symmetry dominates.

Vernadsky's treatment of the problem of the real meaning of time within biological systems is also complex. On the basis of the notion of the specific structure of physical space, part of which is at the disposal of every organism, it could be possible to think about some specific state of time within this unity of "space-time" or "biological time," as Vernadsky called it. This is the so-called "proper time" of living organisms, which is related to their nature and to their own biological cycles, which are distinctly different in length. This time has nothing to do with the movement of celestial bodies or with the laws of the atomic nucleus decomposition on which the present-day definition of absolute time is based. This problem stated by Vernadsky has not been worked out either by natural science or by philosophy.

Alexander Yanshin,

Academician, Cofounder,

International V. I. Vernadsky Foundation,

V. P. Emeritus of the Russian Academy of Sciences, Moscow

Vernadsky at
Borovoy
1942

essays on geochemistry

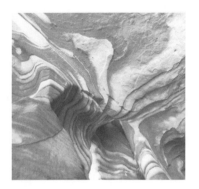

the history of geochemistry

1 geochemistry as a science of the twentieth century

We are living at a turning point in a remarkable era of human history. Events of extreme importance and profundity are taking place in the realm of human thought. Our fundamental views of "the Universe," "Nature," or "the Whole," so much spoken about in the eighteenth and the first half of the nineteenth century, are changing before our eyes with incredible speed. Not only theories and scientific hypotheses, those ephemeral products of intellect, but also new exact empirical facts and generalizations of exclusive value make us rebuild and reconstruct the picture of nature that has remained untouched and almost unchanged by many generations of scientists and thinkers.

The new worldview, in fact the profound renewal of the centuries-old ideas about our surroundings and ourselves, captures us more and more every day. It inevitably penetrates into the domains of separate sciences, into the field of scientific work. These new views concern not only the inert matter that surrounds us, but they also embrace the phenomena of life; they change our notions in the fields of knowledge that we consider the closest to us and the most important. We can say that never in the history of human thought, have the idea and feeling of the whole, and of the causal interrelation of all observed phenomena, possessed such depth, sharpness and clarity as they have reached now in the twentieth century.

The study of the change that has taken place in these ideas and notions makes us think that we are still very far from the ultimate result, and that we have hardly begun to realize the route that the new scientific work has taken. This must be taken into account while evaluating the new concepts of atoms and chemical elements that penetrate our present-day science. They are taking shape in the unsteady, changing, and almost unfamiliar picture of the cosmos. Atoms and elements – the old intuitions of ancient thought – are constantly changing their image and taking new forms in these new and contradictory circumstances.

Each chemical element, we think, corresponds to a certain atom or atoms that are distinctly different in composition from other atoms corresponding to other chemical elements. The atom of the science of the twentieth century is not the atom of the ancient thinkers of Hellas and India, and neither is it the atom seen by the Moslem mystics of the Middle Ages and by the scientists of the last four centuries of our civilization. It is quite a new idea – a new notion. And although the historical roots of our present-day thoughts can be traced back to the atoms and elements of ancient science and philosophy, the changes undergone by them are so great that there is nothing left from the old concepts except the names.

Everything has changed crucially. Maybe it would have been more correct to give a new name to the "atom" of the twentieth century, because it could be done without doing any harm to the historical truth. Our atom does not in the least resemble the matter it forms. The laws of its existence are not similar to the laws of the matter formed by it. In matter, in its physical and chemical characteristics, we observe only general statistical manifestations of large conglomerates of atoms. These show, in a vague and complicated form, only an insignificant part of the characteristics of the atoms themselves, and of their inner structure.

A deep gap separates the scientific model of our surroundings and ourselves – according to the manifestations conditioned by our senses (the macroscopic view of the cosmos) – from the scientifically constructed cosmos where the atom reigns (the microscopic view of the cosmos). The principal physical notions, as well as the method of scientific thinking, suffer a crucial change in these manifestations. As soon as we make scientific advances into the world of the atom, our concept of physical causality sharply changes and deepens, while the century-old ideas about it are destroyed. [In addition to the micro and macro concepts of matter] a third aspect of the cosmos is taking shape at present thanks to the success of astronomical observation and research of the twentieth century: the world of space-time scientifically embraced by large numbers, that – like the atom world – cannot be measured by our senses.

These three concepts about the world, about the reality covered by science, are not coordinated. Everything is in a state of creative motion, both scientifically and philosophically. The atom and its corresponding chemical element are present in all three forms of the worldview. It seems very probable that with attempts at further generalization, great significance will be gained through the trend of scientific synthesis that was put forward in the middle of the eighteenth century by the great Serbo-Croatian thinker, Ruggiero Boscovich (1711–1787), and that is drawing more and more attention at present.

An atom is not a formless and structureless "center of forces," but a regular conglomerate that according to Boscovich is comprised of matter and universe.[1] The history of this trend of thought, which seems

1. R. J. Boscovich, *Philosophiae Naturalis Theoria*, Vienna, 1758. There was a series of editions in the eighteenth century. In 1922, a reprint was published in Chicago and in London (Open Court), with an English translation by C. Child.

to outline and anticipate the way of future scientific thought, has not yet been written. Another great natural scientist, Boscovich's contemporary, James Hutton (1726–1797), approached the same world view independently and laid the foundations for contemporary geology.

An integrated scientific worldview does not yet exist, but the countless new facts unveiling the structure of nature in all its aspects make our thought go deeper and deeper into the realm of atoms, and still further, to the minute entities of which the atoms consist; the real units of time and space. These facts have led to the creation of new scientific disciplines that differ from the former ones, which studied matter – the conglomerate of countless atoms – from a statistical standpoint.

In the twentieth century we are witnessing the flowering of this new kind of science of individual atoms in the form of atomic physics, radiology, radiochemistry, and most recently geochemistry – a small part of astrophysics. Geochemistry deals with the scientific study of chemical elements, i.e., the atoms of the Earth's crust and the whole planet. It studies their history, their distribution and motion in space and time, and their 'genetic' correlations on our planet. It is distinctly different from mineralogy, which studies in the same time and space of Earth's history only the history of atomic compounds, molecules, and crystals. In this strictly limited terrestrial planetary field, geochemistry discovers phenomena and laws whose existence we could only anticipate in the boundless fields of celestial space.

It is obvious to us now that the chemical elements are not distributed chaotically in conglomerates of matter in spaces such as nebulae, stars, planets, atomic clouds, and cosmic debris. Their distribution depends on the structure of their atoms. The atomic geometry of space and time, expressed by the history and distribution of atoms throughout the whole length and duration of the cosmos, exists in large and small forms, in the structure of both a cosmic nebula and a minute organism.[2] The same laws regulate great celestial bodies and planetary systems, as well as the smallest molecules and maybe even the more restricted areas of the separate atoms.

More than two and a half centuries ago, the Dutchman, Christiaan Huygens[3] (1629–1695) – one of the greatest scientists – discovered the inevitable identity of matter and the forces of the Universe, and the manifestations of life throughout its entirety. The identity of matter and forces was based on the gravity laws of his contemporary, I. Newton. It embraced also the Cartesian philosophy that reigned supreme in physicists' minds and hindered the understanding of Newton's scientific discoveries and generalizations of 1676 up until 1730–1740. In the seventeenth century, the notion of the unity of, speaking in modern terms, matter and energy throughout the entire cosmos, the whole of space and time, that had sometimes sprung up in the course of centuries, became part of the scientific understanding of the Universe. But Huygens was one of the few scientists who had clearly expressed the inevitable consequence of this notion: the cosmic unity of life we study in the biosphere.

---

2. V. Vernadsky, Izv. Acad. Nauk, *Section of Physics and Natural Sciences*, p. 512 and further. L., 1932.

3. In the paper he had completed a few weeks before his death, "Kosmotheoros," Huygens puts forward his most intimate thoughts about the world's structure, which are the foundation of all his research work. Two principles of immense importance are expressed there very clearly: the identity of the material composition and forces in the whole cosmos and the notion of life as something quite different from inert matter and as a cosmic phenomenon. ("Kosmotheoros," p. 17 and further, Ed. (2), Er. et L., 1704). Huygens had taken into account Redi's principle, the significance of which he had recognized completely. [*Ed.* "Life comes only from life."]

One hundred fifty years after Huygens, the Englishman W. Hyggins, through scientific experiment and observation by spectrum analysis, proved the identity of chemical elements (atoms) of the stellar worlds based on terrestrial manifestations. The present-day creative explosion of ideas has not shattered this essential principle. He expressed it in the new concept of the identity of the basic elements (electrons, neutrons, protons, and the newly discovered positive electrons, or 'positrons'), which make up atoms or chemical elements, and also in that of the genetic, though complicated connection existing between the atoms of different structures. Studying the laws and regularities of the history of elements of our planet, and studying the structure of the Earth's atoms, we study at the same time the regularities of the smallest spaces and smallest moments that are indivisibly connected with the great whole of the cosmos. There are deep analogies between them and even more than just analogies.

Protons, electrons, positrons, photons, and quanta embrace the whole of time and space – all three aspects of the cosmos. They also constitute and embrace atoms. But chemical manifestations of atoms studied in geochemistry are only a small part of the phenomena connected with these main elements of the cosmos. The chemistry of the cosmos and geochemistry, or the atomic chemistry of the planet in space and time, is a small though important part of the reality studied by science. But we must remember and mention at once that the material substratum of space and time is not determined by chemical phenomena and the chemical characteristics of atoms.

2 forms of existence of chemical elements

Geochemistry, or the history of the chemical elements of our planet, could appear only after the new notions about the atomic and chemical elements had come into being. It could appear only recently, but it is rooted deeply in the history of science. Now we can see how the separate studies of diffcrent scientists of the past, which were not quite clear to their contemporaries, appear in a new form under the influence of the great scientific generalizations of the present – how they are receiving a new meaning and prove to be interconnected. Unfortunately, I cannot dwell upon the history of these ideas and upon the rise of geochemistry in detail. The preliminary work has not been done yet, and no full and coherent account can be given here of the way human thought has developed in this field.

No doubt, in the seventeenth century and earlier, systematic studies of geochemical problems were undertaken. A future historian of science will discover a fruitful scientific trend, list the names, trace a series of discoveries, observations and facts that are getting more and more precise, and find the roots of the most significant contemporary empirical generalizations and scientific ideas. This trend had become especially powerful and important by the end of the seventeenth century. Here I shall mention the name of the man who probably realized more of the scope and the importance of the phenomena encompassed by modern geochemistry than anybody else. This was Robert Boyle (1627–1691), a founder of the theory of chemical elements and a creator of modern chemistry.

The history of natural waters and of the world's ocean in particular, the atmosphere as a weighty gaseous medium, the solution of gases in water, the first exact delineation of chemical elements in terrestrial bodies, and the beginning of precise chemical analysis of terrestrial products have all originated from Boyle's works. I cannot, however, dwell upon these and other forgotten scientific studies, which have not actually passed unnoticed, and which have had a certain influence on the contemporary scientific mind. They lasted for almost two centuries. As early as in the second half of the eighteenth century, geochemical problems began to arouse scientific interest, although the idea of a chemical element was vague and far from the notions of the nineteenth and twentieth centuries.

G. F. Rouelle senior (1703–1770), and his even greater junior contemporary, L. Lavoisier (1743–1794), whose creative work was stopped while fully blossoming, had already posed these problems clearly enough. We cannot imagine the heights Lavoisier[4] could have reached. Before his death he began to approach the deepest geochemical problems in his works concerning water and the physiology of breath [respiration]. Rouelle – the senior – published very little, but he influenced his contemporaries greatly by his public experimental lectures on chemistry that he delivered in Paris at the Royal Botanical Garden. All the intellectuals of Paris, or even of Europe, gathered for these lectures, and numerous foreigners who were in Paris – often great minds – attended these lectures. Nowadays it is very difficult to realize Rouelle's influence; nobody has even tried to do it. But it is indubitably enormous, as it spread

---

4. L. Lavoisier, *Traité de Chimie* (1787), Oeuvres I, p. 325, P., 1868; *Opuscules Physiques et Chimiques*, ib. I, p. 445; Ouevres II, p. 870, P., 1872.

from Paris throughout Europe and survived his death. In the works of Lavoisier on the history of elementary gases and on the history of water, there are shining examples of geochemical generalizations expressed in the scientific language we are used to. Due to the great influence of Lavoisier's ideas on the whole of modern chemistry, geochemical problems were introduced as well. Since then, some of these problems have begun to be included in chemistry courses.

His elder contemporary, Leclerc de Buffon (1707–1788), who was still very far from our present-day notion of chemical elements, presented in his history of minerals a series of brilliant and interesting generalizations and posed a series of significant geochemical problems. He was able to do this not only because he was a profound observer of nature who covered all the scientific knowledge of his time, but also because he lived in the midst of social activities and was an agronomist and a technologist. We find geochemical problems in his chapters on the history of native elements, and on that of metals in particular. But we find them also in other parts of his *Les Epoques de la Nature* (1780) and *Histoire Naturelle, Générale et Particulière* (1749 and following years). Not only was Buffon a great writer, he was also one of the greatest and most profound naturalists, one of the few people who had indeed observed the Universe as a whole. In these "Essays" we shall come across his ideas and their consequences more than once.

We cannot but mention also M. V. Lomonosov (1711–1765), another contemporary of Rouelle and Buffon. Only nowadays have we fully appreciated Lomonosov's scientific thought, which foresaw the future ways of science. In his forgotten works, which have been published badly and incompletely, his understanding of geochemical problems can be seen clearly and distinctly. In the Petersburg Academy of Sciences he had followed his own path, onto which scientific thought arrived only in the twentieth century. Incessantly he went deeply into the chemistry of natural bodies in general and in connection with Earth's history. By the beginning of the twentieth century, the new chemistry had been integrated into scientific thought: the chemical element had received a new understanding, far from that attached to it by Lavoisier. Geochemistry had seemed to be on the verge of appearing, but it was created much later. Apparently, the empirical material had not yet been sufficient and understanding of the chemical element itself was not clear enough.

It was the time of the creation of present-day chemistry and geology, and their synthesis gave birth to geochemistry. At the same time, applied scientific disciplines were being created on the basis of the new scientific ideas of matter and our planet – they were understood as technology then. This is extremely important too, because both in 'technology' in the broader sense and in pure knowledge we come across thoughts about geochemical problems – the deepening of these problems. The history of chemical elements in the Earth's crust, their role in different chemical processes, and particularly in the phenomena of life – both in living nature and in daily human life – has permeated scientific thought since the end of the eighteenth, and beginning of the nineteenth century; it has had various manifestations occurring everywhere.

I shall mention three of the most outstanding predecessors of modern geochemistry of the last century. These are the Englishman Humphrey Davy (1778–1829), the Prussian German J. C. Reil (1759–1813), and A. von

Humboldt (1769–1859). The brilliant key work of A. von Humboldt, a Prussian by origin, was published outside Prussia. As for himself, in the first decade of the nineteenth century he was completely under the influence of the intellectual atmosphere of Paris.

Humphrey Davy was a brilliant experimenter, physicist, and chemist who covered all the science of his time; he was a thinker possessing a deep poetic understanding of nature. He always connected science with life and was one of the most brilliant figures of the first half of the nineteenth century, which was so rich in talented people. Davy made a tremendous impact upon the science of his time by his lectures, numerous articles and books, and by brilliant experiments. In his works we find a lot of data about the history of chemical elements in the Earth's crust. In this field he developed the ways discovered by Rouelle and Lomonosov on a new scale. His works were a prototype of all the later treatises of chemistry in which the account of the properties of chemical elements is always connected with their geochemistry. In the later works of Dumas, Bercelius, Liebig, Mendeleyev, and other, no less talented scientists, we always find speculations or brilliant generalizations concerning geochemical problems. After Davy, during the entire nineteenth century, geochemical problems were included into inorganic chemistry courses; they were studied while discussing particular chemical elements.

The fate of Reil was quite different. One of the most outstanding doctors of his day, absolutely committed to helping the suffering, he did not spare himself and died on duty. Reil died in the very midst of his scientific searches. Being a doctor, an anatomist, a psychiatrist, and a physiologist, he was not interested in geochemical problems directly. But he was a man of broad philosophical thinking, a naturalist, like all the genuine doctors of his day. As a philosopher he shared the trends of natural philosophy, and apparently he was close to Schelling, but his thought was independent. His contribution to the history of geochemistry is connected with the study of *the chemistry of organisms.* He was the first in the era of the new chemistry to suggest the importance of the chemistry of organisms, and in this respect he was far ahead of his time.

The roots of Reil's aspirations and ideas go far back into the medical tradition. Beginning with the petrochemists of the seventeenth and eighteenth centuries, maybe even with Paracelsus (Bombast von Hohenheim, 1493–1541), the importance of chemistry in medical systems and in the understanding of healing the sick had never left the intellectual horizon of doctors. Generations of doctor-chemists follow one another incessantly for centuries. Reil considered the thorough, quantitative chemical study of organisms necessary, and he searched there for the answer to the manifestations of life. He was an innovator whose work was stopped by death at the very beginning. It is difficult to say what Reil's contribution could have been, had his life been longer.

This was also the way of thinking of one of the most striking people of the first half of the nineteenth century: Alexander von Humboldt. In his early works, especially in "Flora Fribergensis Specimen" (1793), written before he plunged into South American nature, A. von Humboldt had come very close to many of the present-day problems of geochemistry. These studies of the young Humboldt were interrupted by his long journey, the processing of its results and the creation of the striking synthesis presented in his *Cosmos.* As an old man, in the fifth volume of *Cosmos,* he returned to one of the geochemical problems: the influence of life on its surroundings. But death stopped this work in the middle of a word.

In the paper of 1793 mentioned above, there was a brilliant effort to describe living organisms from the point of view of their chemical elements; being a mineralogist and a geologist, Humboldt never ceased seeking for their origins in the inert matter surrounding the plants. Decades passed until the problem was posed again as clearly as it had been by Humboldt. His way of putting forward the problem of the geographical spreading of organisms goes far beyond the limits of the studies of his followers; and deeper than the new branches of geography that appeared under its influence; it approaches the geochemical concepts of our days. He considered living matter to be an unbreakable and regular part of the planet's surface, inseparable from its chemical environment.

During the entire nineteenth century, the field of contemporary geochemistry was being prepared. Step by step, the picture of the unity of the chemical composition of the Universe was becoming clear. This unity was first put on an experimental basis after the idea of the cosmic origin of meteorites penetrated the scientific mind. That idea was born in the first quarter of the nineteenth century, thanks largely to the continuous (1794–1826) scientific work of E. F. Chladni (1756–1827), an original scientist who, like Humboldt, stood apart from German university science. Chladni, who was not a chemist, followed his own path in life and was an innovator in science. The chemical composition of meteorites being identical to that of *terrestrial bodies* was first stated by E. C. Howard (1802), and at the same time J. L. de Bournon found out how they differ *mineralogically*. Both statements soon entered the scientific mind, but conclusions were drawn much later.

The notion that the chemical elements of living organisms were identical to those of inert matter was slowly acknowledged by science. Until the 1740s, it was not considered scientifically proven and was checked by special experiments. By the middle of the nineteenth century, following the methods brilliantly worked out by H. Davy, scientists had discovered the principal features of plant nutrition, which were then immediately taken up on a planetary scale (i.e., studied not only in their biological, but also in their geochemical aspects). This tradition has continued since the time of Lavoisier.

J. B. Dumas (1800–1884), J. Boussingault (1802–1887), K. Sprengel (1787–1859), J. von Liebig (1803–1873) and many researchers who followed them, or their contemporaries whose work was partly independent, stated the geochemical significance of green life. As we shall see, this life refers to the main part of the living matter of the biosphere. Dumas, Boussingault, and Liebig discovered the importance of green life in the gas exchange of the planet, and apparently it was Boussingault who had the deepest understanding of it, for he understood the geochemical aspect of the phenomenon best of all. He came across it outside the laboratories – in nature – during his long stay in the tropics and his studies of volcanic phenomena and minerals. In this field he was one of the shrewdest thinkers of the nineteenth century, and up till now we find in his works new material that has not been covered by scientific thought yet. Sprengel and Liebig furthermore discovered the real significance of the *ashy elements*. The theoretical constructions of Liebig influenced our understanding of these phenomena and completely reversed the explanation of the century-long characteristic of human culture – the importance of fertilizers for the productivity of soils. They also showed the geochemical role of green plants by using the compounds of phosphorus (this was clear to Boussingault too), potassium, and other elements needed by the plant.[5]

At the same time chemists did other studies on the minerals, waters, gases, and rocks surrounding us. Many scientists, especially chemists, considered mineralogy to be "the chemistry of the Earth's crust," as I. J. Bercelius called it. Gradually, precise research on the nature of minerals amassed an enormous amount of material. At the same time, by the end of the nineteenth century, the chemical analysis of rock formations, the research on waters, and the chemical study of fossil minerals gave a solid basis for empirical generalizations, for the creation of biochemistry.

Now we can see (it was remembered in 1931–1932) that different people understood the process being in progress quite clearly, and that the notion and the word "geochemistry" had already been created by that time. It was done in the 1830s and early 1840s by an original scientist from Basel – C. F. Schoenbein (1799–1868). His ideas were forgotten, but a historian of thought cannot forget the real influence of such an outstanding and brilliant personality as Schoenbein, who discovered ozone and worked in his own peculiar way. Schoenbein was not alone; he exerted a great influence on his surroundings. His articles

---

5. I was surprised that Vernadsky did not include the German scientist Julius von Sachs (1832–1897) in this discussion. Around 1850, von Sachs (sometimes referred to as "the father of plant physiology"), basing his studies on those of the scientists Vernadsky does mention, developed hydroponic methods (plants grown in solutions of minerals) to study the elements required by plants, especially the so-called *major elements*: K. Ca, Mg, N, P, and S (in addition to the H, O, and C of water and carbon dioxide). "Minor elements" were later shown to be essential: Cl, Fe, B, Mn, Zn, Cu, and Mo. These (and other elements required for some species) were apparently present as impurities in the solutions of von Sachs. (W. Knop was also involved in this work.) –Ed.

and, as we can see now, his letters, are full of ideas and research that went beyond the limits of the science of his day; these were partly echoes of the past and partly anticipations of the future. Apparently, a friend of Schoenbein's, M. Faraday (1771–1867), was not indifferent to his geochemical interests; his life was closely connected with that of Humphrey Davy whose significance for geochemistry I have already mentioned.

In 1842 Schoenbein wrote: "A few years ago I had already put forward the idea that we must have *geochemistry* before speaking about a real geological science that pays at least as much attention to the chemical nature of the matter comprising the Earth, and to their origin, as to the relative antiquity of these formations and the fossils of antediluvian plants and animals buried there. Of course, we can be sure that the geologists will at last cease to follow the trend they are supporting now. In order to broaden the limits of their science, they will have to look for new auxiliary means as soon as fossils fail to satisfy them. They will undoubtedly introduce mineralogy and chemistry into geology then. The time for this to happen does not seem to me very far from now. . . ."[6] Now these words seem prophetic to us. Schoenbein was mistaken regarding one point though: the time for his ideas came only in the twentieth century, decades after his death; then the word he had created was reborn and embodied by a new geological science.

By 1850, namely during the period of 1847–1849, brilliant and outstanding geochemical generalizations were published in scientific works that had collected an enormous amount of exact facts,

---

6. C. F. Schoenbein. Mitteil. aus d. *Reisebuche eines Deutsches Naturfforschers*, S., 99. Basel 1842

which thereby entered general scientific thought and influenced it. This was done by three prominent naturalists who worked independently from each other, and whose works complement one another. None of them could cover the whole field of geochemistry by one synthesis but, as we see now, the results of their extended works which appeared almost simultaneously, presented a general outline for our new science. Nevertheless, their contemporaries did not see it – they noticed only contradictions and could not perceive them as part of one and the same whole.

These three naturalists were: Prof. K. Bischof (Bonn), who published in 1847 the first volume of his *Lehrbuch der Chemischen und Physikalischen Geologie*;[7] Prof. Elie de Beaumont (Paris) who published in *Bulletin de le Societe Geologique de Paris* a brilliant memoir about volcanic phenomena,[8] which was not understood by his contemporaries; and Prof. J. Breithaupt (Freiberg), who synthesized in 1849 the century-long work of the Freiberg mineralogy school in his book *Paragenesis der Mineralen*.[9] In these works we already have clear and solid roots for the main data of geochemistry. If somebody at that time, for instance in 1850, could have embraced all that material at once, we would have had geochemistry in the nineteenth century. Nevertheless, it was formed only in the twentieth century.

No one was able to embrace all this material due to the peculiar atmosphere of geological work at that time. It was the time of the

---

7. K. G. H. Bischof. *Lehrb. d. Chem. u. Phys. Geol.*, 1–2. Bonn, 1847–1855

8 Elie de Beaumont, "Bull. d. 1 Soc. Geol. d. Fr." (2), 4, p. 1249, P., 1846

9. J. Breithaupt, *Paragenesis d. Mineralen*, Fr., 1849

argument between neptunists and plutonists, which was dying away but had not been finished yet, and which had involved three generations of scientists in the eighteenth and nineteenth centuries. One party, the neptunists, considered surrounding terrestrial nature to have been created by the forces of water and formed at normal temperature and pressure. Life, which was closely connected to water, occupied its honorable place in the creation of nature. According to the neptunists, life was a great force, not an accidental phenomenon in the history of the planet. The other party – the plutonists – paid no special attention to the forces and phenomena of the Earth's surface. They believed that the great forces inside the planet, which they thought to be still in a state of incandescent magma, were creating the nature of the Earth. Life, in all its variety and apparent importance, was just an insignificant peculiarity that did not reflect the main phenomena of the planet. The forces, whose activity manifested themselves in volcanoes, geysers, earthquakes, and thermal springs, formed all the principal features of the Earth's surface and influenced the formation of mountains, rocks, and conglomerations of water and gases.

These two opposite concepts of our planet really concerned the main features of a worldview. The choice between them, once taken, led to opposite conclusions, which had great vital significance for the importance of life in the structure of the cosmos. The meaning of these old arguments, in the mental life of that time, can be understood from the creative work of the great naturalist and poet – a brilliant and passionate neptunist – J. W. Goethe. The second volume of his "Faust," which embodied his lifelong efforts to express his concepts of the future and the tasks of human life, is permeated by reflections and echoes of this argument.

K. G. H. Bischof (1792–1870) became a neptunist having realized the significance of the Earth's surface for the history of the planet through long speculation and experimentation; in the early years of his scientific life he had been a plutonist. This revolution in his views affected his whole work. He proved the importance of water, collected an enormous quantity of facts, gave clear pictures of the history of many chemical elements, and eventually showed that in the phenomena of inert matter, their history could be reduced to cyclical processes that, in the first part of his paper, he considered a typical feature of organogenic elements. For organogenic elements this picture had already been given by Dumas, Boussingault, and Liebig. In this connection, the phenomena of life in the chemical processes of the Earth were put at the forefront in his paper. The influence of his work was immense not only on the continent, but also in English-speaking countries; Bischof himself was connected with English scientific circles.

Unlike Bischof, Elie de Beaumont (1798–1874) was a plutonist, who put forward that the connections between chemical elements and the regulations of their locations are consequences of magmatic and volcanic processes. For a long time, the brilliant work of Elie de Beaumont attracted little attention outside France, partly because of the domination of neptunic ideas and partly due to his unsuccessful hypotheses about the formation of mountain chains. But long after his death the truth of his generalizations was confirmed by exact observation, and became an indispensable part of geochemical work.

The exact empiricist, J. Breithaupt (1791–1873), also followed an independent route. Using the experience of mining, he put forward correlations between elements that are situated together – and which went beyond the schemes of pure neptunists and plutonists. The processes studied by Breithaupt did not fit into their simple schemes, and he discovered new phenomena of our planet that had been one-sidedly described both by Elie de Beaumont and by Bischof. Breithaupt was not alone; exact empiricists, observers of ore deposits amongst whom were both neptunists and plutonists, were following the same route at that time. The most outstanding investigations were those of J. Durocher (1817–1860), J. Fournet (1801–1870) and W. Hennwood. New properties of water were found, and the influence of the high temperature of lower geospheres became clear. The investigation of these processes from the standpoint of ore deposits – mainly metals – inevitably made the scientists study the history of chemical elements in the Earth's crust.

As both plutonic and neptunic schemes were disappearing, the scientific work of the second half of the nineteenth century continued in all these directions. Geology soon left the old schemes and covered the complexity of nature with more diverse theories. At the same time, the chemical mind was distracted from geochemical problems; in the history of chemical elements much attention was being paid to properties that did not seem to manifest themselves in the processes of our planet. The idea of a chemical element became more abstract, it seemed that there was an insurmountable barrier between chemical and geological sciences. This was clearly shown in the different classifications of sciences that were so numerous at that time. The state of mind of researchers was unfavorable for creating geochemistry.

The generalizing and deep view of chemistry that brilliantly combined the traditions of Rouelle, Lavoisier, Davy, and Bercelius, and that was interpreted by such an original and powerful mind as D. I. Mendeleyev in his *Foundations of Chemistry*, stood absolutely alone. In *Foundations of Chemistry*, the problems of geochemistry and space chemistry were not only fully described, but were also often dominant. As always with D. I. Mendeleyev, it was not a repetition of someone else's materials, but it contained something new, something found by his brilliant personality, grasped by his shrewd mind.

In general, neither in geochemistry nor in chemistry did a favorable environment exist for the development of geochemical problems into an integral, separate, and scientifically based new discipline. The soil had not been ready, and it was slowly being prepared for decades, beginning in the second half of the nineteenth century. There were three changes in the ideas about the environment that provided a solid basis for this new science in the twentieth century.

In the second half of the nineteenth century our notions about the chemistry of the cosmos began to change. The unity of its chemical composition, which – as we could see – had been clear to Huygens in the seventeenth century and had been confirmed by the analysis of meteorites, received a new and solid affirmation in 1859 with the discovery of spectral analysis by G. R. Kirchhoff (1824–1887) and R. Bunsen (1811–1899).[10] This discovery expanded the human horizon enormously. In fact, it was one of the deepest insights into the structure of matter; spectral analysis proved the chemical unity of the universe. But at the same time, thanks mainly to spectral analysis and to the development of our notions about the complicated unity of matter expressed in its atomic aspect, which led to a deeper theoretical understanding of the great scientific generalization of the Periodic System of elements, thanks to all this, the very notion of the *chemical unity* of the world became enormously deeper and wider.

---

10. G. Kirchhoff and R. Bunsen, "Annalen d. Phys. U. Chem.," 110 u. 111, I. , 1904.

On the one hand it became clear that the atoms of our planet were present in different states. It was also necessary to admit the existence of certain states of atoms (i.e., chemical elements of the universe) that cannot exist on planets including the Earth.[11] On the other hand a question arose whether the atomic manifestation of matter – its *chemical* composition – corresponds to the dominating mass of matter dispersed throughout the time and space of reality. The spectral analysis in the works of Kirchhoff and Bunsen clearly and definitely confirmed the existence of chemical elements in dispersion – all matter of the Earth being permeated by them. For some elements such as sodium this was already understood by H. Davy and then by others, but this notion entered the general scientific mind only after the works of Kirchhoff and Bunsen. Nevertheless, the idea of its importance for geochemical problems was put forward only in the twentieth century (1910). Up until now the phenomenon has not been completely covered by scientific thought and even less by experiment.[12]

Apparently there is not only one form of planetary atoms in a specific state. It is clear that for some chemical elements, for instance lead, isotopic mixtures can differ. This is caused by radioactive dissociation and specific conditions of the atoms' migration. It is possible that there is another phenomenon related to it such as the influence of life – the change of isotopic mixtures in the biosphere – but this issue is not quite clear yet. Eventually it was understood

---

11. Perhaps Vernadsky was thinking of the "state" of a super-heated ionized gas (a plasma) as in the solar corona – or in a fluorescent lamp, which now exists on Earth! —*Ed.*

12. Cf. V. Vernadsky, *The Biosphere.*

that geochemical problems made up an inseparable part of the problems of cosmic chemistry, that the chemistry of Earth was one of the manifestations of planetary chemistry, and that the theory of the geochemical character of chemical elements, i.e., geochemistry, was distinctly different from *mineralogy*, the study of molecules and crystals formed by atoms.

Our idea of the unity of the chemical composition of the universe undergoes still deeper changes under the influence of the growing understanding of the fact that the atom is not the dominant form of manifestation of matter in the universe. Studying atoms gives no definite idea about the matter of the cosmos. Beyond atoms we can observe the realm of electrons, positrons, neutrons, free protons, and a series of unknown material particles dominating in mass. To a lesser degree these phenomena also cover the matter of the Earth, for instance the electrons of the ionosphere's electric field. The electronic chemistry of general chemical physics must occupy the dominant place in cosmic chemistry and must take its place in the chemistry of our planet together with geochemistry and mineralogy.

Since the last century, considerable changes have taken place in our scientific notions about the research area of geochemistry and mineralogy – the geological substratum of the planet. In the first half of the nineteenth century. it was considered indubitable that geological phenomena could serve as the basis for conclusions concerning *the whole Earth*. The dispute between plutonists and neptunists was based entirely on this assumption. The thin surface film, our biosphere filled with life, seemed to get lost and forgotten in the mass of the planet. Slowly but steadily, from generation to generation, these notions disappeared, because it was gradually discovered that

all the geologically studied processes relate only to the outer part of
the planet, the Earth's crust. Processes that had formerly been related
to the inside of the Earth proved to be external.

Gradually the boundaries of the Earth's crust were determined;
they did not exceed the upper 100 kilometers. Geodesists were the
first to take up this viewpoint, and as early as 1851 the English priest
J. G. Pratt provided the foundations of the theory of isostasy; that
is, the non-homogeneous structure of the outer part of the planet,
the Earth's crust, as opposed to the homogeneous structure of the
deep layers of the planet. He pointed out that it was not the depth
of the Earth but the Earth's crust that was involved with the greatest
phenomena we know on the surface of the Earth: the formation of
mountain chains. Immediately, the English astronomer G. B. Airy
(1855) expressed Pratt's ideas more correctly, and explained them
by the hydrostatic equilibrium of different heterogeneous parts of
the Earth's crust. Pratt's ideas, formulated in a general form by the
American geologist C. E. Detton thirty years later, gained scientific
acknowledgment only in the twentieth century. Geologists, however,
came to the same conclusions earlier and in a different way. Thanks
to that insight, all views on geochemical problems changed clearly.
The volcanic products, the products of life, and the sediments of the
sea proved to be bodies of one and the same planetary field that, as
well as the phenomena of life, is different from the large mass of the
Earth. As a general geochemical understanding, the significance of
life increased and changed essentially.

In that period of time, another revolution in our general worldview
was taking place. The old idea of J. Dalton and W. Wollaston, its logical
consequences which were perhaps not quite clear to themselves,
became reality; *the atom and the chemical element proved to be identi-*

cal. In order to understand the atom, one had to study the chemical element. The atom became as real to us as the chemical element; it acquired flesh and blood and became a real body. This achievement of science took place in the twentieth century, but the late decades of the previous century, in spite of what contemporaries thought, were already leading the scientific mind toward this generalization. It is well known that by the end of the nineteenth century, the atomistic view of Earth's environment seemed to be losing ground and was being replaced by dynamic ideas about the world. In reality it was a mirage; in reality the atomistic view has never been as influential in the scientific worldview as today. True, the atom in the new worldview has little to do with the atom of philosophers and even of physicists; it is the chemical element of chemists in the form of an atom.

All these changes had for the first time made it possible to embrace geochemical problems as a whole, as a special scientific discipline, and to separate geochemistry out as *a science that studies the history of atoms* (understood as the chemical elements on our planet). Actually, we are studying only its external envelope, the Earth's crust. In particular, this separation of the new science was taking place more or less independently in different parts of the civilized world. In Washington, F. Clarke – a chemist from the American Geological Committee who had studied geological problems all his life – collected and arranged an enormous amount of material in his book, *Data of Geochemistry,* of which the first edition was published in 1908. This book exerted a great influence on scientific thought and was published in five editions (the last one in 1924).[13]

---

13. F. Clarke. *Data of Geochemistry.* W., 1980; fifth ed., W., 1924. On Clarke cf. "Nature," 120, p. 214. L., 1931; W. Schaller. *Americ. Mineraologist,* 16, 405, Men., 1931.

A huge amount of factual data is collected in this book. Clarke tries to give the exact numerical data concerning the history of the main chemical elements. Although in his youth (1872) he had been one of the first scientists who had dared to scientifically tackle the issue of the possibility of turning one chemical element into another in connection with their history in the cosmos, and although fifty-three years later he returned to these cosmogonic generalizations, in his *Data of Geochemistry,* he pursued not hypotheses and wide generalizations, but comparison and criticism of exact numerical data on the history of chemical elements in the Earth's crust and in its processes. He was interested in the study of the composition of the sea, the average composition of rivers, and the study of the Earth's crust; everywhere he introduced new numbers and critically revised the old ones.

Clarke's book has in fact become the foundation for further generalizations and further geochemical work.[14] It summed up and covered a tremendous amount of material connected with the numerically exact chemical, geological, and mining research on the American continent. At the same time, an American who had worked in Canada, his elder contemporary T. Sterry Hunt (1825–1892), was also attempting "the synthesis of the Earth," as he put it. Sterry Hunt's influence was great, but he left a lot of room for theoretical speculations which were not always successful. Clarke's synthesis, which was being built at the same time and on a firm empirical basis, proved to be more solid.

---

14. In spite of the great merits of this book, one cannot but notice some significant gaps that are especially conspicuous in the latest editions. For instance, the data on soils are ignored completely. Hydrogen sulfide was not covered at all, though its geochemical importance is enormous. The composition of living matter apart from the skeleton is not taken into account. Finally, the non-American post-war literature is incompletely referred to.

Having collected the facts and having empirically generalized them into the new science of geochemistry, Clarke finished Bischof's work in the twentieth century. His book gave a summary of the tremendous work of thousands of people over a long period of time. As early as 1882, his first calculations of the gross chemical composition of the Earth's crust had appeared. After that, Clarke incessantly altered and improved them (for the last time in 1924, together with H. Washington). These data – Clarke's numbers – did not influence the scientific mind for many years but were objected to, and were appreciated for their great significance only in the last decade. This significance may turn out to be even greater than Clarke thought if the resemblance of the outer envelope of our planet to the outer envelopes of other planets can be proven.

As we shall see below, Clarke followed the routes outlined by W. Phillips as early as in the beginning of the nineteenth century, and he was the first to seek not an approximate numerical estimation of the phenomena, but a concrete exact number. Clarke did not formulate the task of geochemistry distinctly and categorically as being the study of the history of the planet's atoms, this trend in geochemistry appeared later and aside from his direction of thought. But thanks to the real significance of Clarke's numbers in the new theories of atoms, to their influence on the physical and chemical thought of the twentieth century, his work has completely entered notions that were forming outside the realm of his thought. His geochemistry corresponded to the chemical and physical geology of Bischof but it met a different scientific environment.

The notion of geochemistry as a science about the history of terrestrial atoms appeared as a background to the new atomistics, chemistry and physics, in close connection with the idea of mineralogy typical of the Moscow University in 1890–1911. Both in teaching and in scientific mineralogical work there, most attention was paid to the history of minerals – their genesis and their change – which usually occupied a second rank in the mineralogy of schools of higher education at that time. With such a presentation of mineralogy, geochemical problems were presented on a larger scale, and were considered more important[15] than in the common university courses of inorganic chemistry. Gradually, the work of the Mineralogy Chair of the Moscow University, and later the work related to it at the Mineralogy Museum of the Academy of Sciences, was more and more directed toward geochemistry. The name given by Clarke immediately found content here (although different from his own) and fruitful ground. The phenomena of life, and the mineralogy of sedimentary rocks in connection with radioactivity and general issues concerning the properties and character of atoms, occupied a considerable place. In 1912 in Moscow, in the university named after Shaniavsky, A. E. Fersman delivered the first university course of this new science. Furthermore, a series of A. E. Fersman's and Ya. V. Samoylov's works (1870–1925) have firmly established the traditions of geochemical work in our country.

By the twentieth century, the study of ore deposits, which had made great progress by that time, contributed greatly to the creation of geochemistry. The close connection between geochemical problems and insights about ores, which had led to the generalizations of

---

15. I cannot but remember my teacher, V. V. Dokuchaev, whose interest in the origin of minerals was reflected in his lectures and in the talk of young scientists surrounding him. It was on his advice that my article on the origin of minerals appeared in the *Encyclopedic Dictionary*.

Bischof, Breithaupt, and Elie de Beaumont in the previous century, has never been interrupted. But in the new century it acquired quite a different appearance due to the progress of chemistry, the unusual deepening of technology, and the great scope of extracting old metals and introducing new metals into the structure of human economy and life. In our century this phenomenon has acquired an extraordinary form: that of the global economy.

The works of the Frenchman, L. de Launay, and the German, A. Stelzener (1840–1895), were of great influence and posed geochemical problems. But considering the problems of the theory of ore deposits or applied mineralogy, most significant in the creation of the field of geochemistry are the works of the Norwegian, I. Focht (1858–1932), which are closely connected with the century-long mineralogical research based on the nature of Fennoscandia, and the works of North American mining engineers such as C. Van Hise and W. Lindgren. They connected the problems of geochemistry to that of ores, and in this way gave it a great practical value. This applied significance of geochemistry is growing rapidly during recent years. It manifests itself in our country as well, but we have to say that the conditions for its correct development are not favorable here.

Modern geochemistry is closely connected with the work and thought of another scientist – Prof. V. M. Goldschmidt – who in 1930 created the most powerful scientific center of geochemical work in Göttingen, Germany, although he himself was a product of the century-long scientific traditions of the Norwegian school of mineralogy. From 1914 to 1930, V. M. Goldschmidt, disciple of the outstanding mineralogist V. Brögger, was a professor in Christiania (now Oslo), where he created a mineralogical and geochemical institute with a high level of scientific thought. The Institute of Göttingen made further progress. The nature of Fennoscandia gave the mineralogical work in that country quite a unique flavor; it is an area of ancient crystalline rocks, and radioactive minerals are also present.

Often they are quite unusual in beauty and manifestation, distinctly different from all others in their outer form, unique in color, shine, and chemical composition, and also in such physical properties as metamic structures, compounds of uranium and thorium, rare earths, titanium, niobium, tantalum, zirconium, and hafnium. The school of chemists and mineralogists, which was here for centuries, covered this most difficult group of terrestrial bodies and discovered in the native material a quantity of new minerals and new elements. Bercelius, proceeding from this native material, applied his thought and exact methods to the whole area of inorganic chemistry of the twentieth century.

At the end of the century, Brögger synthesized the mineralogical work of the Fennoscandian and German scientists with reference to the same natural bodies. W. C. Brögger (born in 1851), a man of rare knowledge and exactitude of work, is equally prominent in geology, paleontology, and crystallography. He is a first-class researcher both in the field and in the laboratory; he connected the chemical study of minerals with their crystalline structure, developing in this field the ideas of another Norwegian scientist, the chemist and mineralogist T. Hjortdal (1839–1925).

Brögger's disciple, V. M. Goldschmidt, therefore approached geochemical problems in surroundings full of traditions. The deeper geospheres of the Earth's crust that are located beyond the stratisphere and the biosphere drew his attention; they constitute the largest part of the substance of our planet open to research. Solid matter acquired a special significance, and due to a new specification of roentgenometric methods, led to creation of crystal chemistry, in which Goldschmidt played an important role. Working in this direction, and taking into consideration the processes of elements' migrations in the vectorial solid medium, Goldschmidt introduced into geochemistry a notion fraught with many consequences: the notion of the chemical elements' behavior, as conditioned by their structure. He pointed out the regularities of their manifestation in the solid medium forming the Earth's crust. Goldschmidt's Institute in Göttingen is at present the largest center for this kind of scientific work.

Geochemistry is developing rapidly now; its influence and significance in purely scientific issues is constantly growing and increasing. The preparatory period is over. Separate branches are beginning to spring up thanks to the close connection of the large complex of its problems with fields that are in fact separated from geological disciplines, one of which is geochemistry. In this way, biogeochemistry has begun to separate. In 1927, the center for research in this field was established in Russia at the Academy of Sciences: its Biogeochemical Laboratory, which is not very powerful as yet.[16]

---

16. In 1947 this laboratory was turned into the Laboratory of Geochemical Problems, and then in 1947 it became the basis for creating the Academic Institute of Geochemistry and Analytical Chemistry named after Vernadsky. (*Russian Editors.*)

table 1

PERIODIC TABLE OF THE ELEMENTS

Legend: Element Name / Symbol / Number / Atomic Mass

Category blocks: Metals · Metalloids · Nonmetals

Group IA	IIA	IIIB	IVB	VB	VIB	VIIB	VIIIB	VIIIB	VIIIB	IB	IIB	IIIA	IVA	VA	VIA	VIIA	VIII
Hydrogen H 1 / 1.01																	Helium He 2 / 4.00
Lithium Li 3 / 6.94	Beryllium Be 4 / 9.01											Boron B 5 / 10.8	Carbon C 6 / 12.01	Nitrogen N 7 / 14.01	Oxygen O 8 / 16.00	Fluorine F 9 / 18.00	Neon Ne 10 / 20.18
Sodium Na 11 / 23.00	Magnesium Mg 12 / 24.30											Aluminum Al 13 / 26.98	Silicon Si 14 / 28.08	Phosphorus P 15 / 30.97	Sulfur S 16 / 32.07	Chlorine Cl 17 / 35.45	Argon Ar 18 / 39.95
Potassium K 19 / 39.10	Calcium Ca 20 / 40.08	Scandium Sc 21 / 44.96	Titanium Ti 22 / 47.88	Vanadium V 23 / 50.94	Chromium Cr 24 / 51.00	Manganese Mn 25 / 54.94	Iron Fe 26 / 55.85	Cobalt Co 27 / 58.93	Nickel Ni 28 / 58.69	Copper Cu 29 / 63.55	Zinc Zn 30 / 65.39	Gallium Ga 31 / 68.72	Germanium Ge 32 / 72.61	Arsenic As 33	Selenium Se 34 / 78.96	Bromine Br 35 / 79.90	Krypton Kr 36 / 83.80
Rubidium Rb 37 / 85.47	Strontium Sr 38 / 87.62	Yttrium Y 39 / 88.91	Zirconium Zr 40 / 91.22	Niobium Nb 41 / 92.91	Molybdenum Mo 42 / 95.94	Technetium Tc 43 / (98)	Ruthenium Ru 44 / 101.07	Rhodium Rh 45 / 102.91	Palladium Pd 46 / 106.42	Silver Ag 47 / 107.87	Cadmium Cd 48 / 112.41	Indium In 49 / 114.82	Tin Sn 50 / 118.71	Antimony Sb 51 / 121.75	Tellurium Te 52 / 127.60	Iodine I 53 / 126.90	Xenon Xe 54 / 131.29
Cesium Cs 55 / 132.91	Barium Ba 56 / 137.33	Lanthanum *La 57 / 138.91	Hafnium Hf 72 / 178.49	Tantalum Ta 73 / 180.95	Tungsten W 74 / 183.85	Rhenium Re 75 / 186.21	Osmium Os 76 / 190.2	Iridium Ir 77 / 192.22	Platinum Pt 78 / 195.08	Gold Au 79 / 196.97	Mercury Hg 80 / 200.59	Thallium Tl 81 / 204.38	Lead Pb 82 / 207.2	Bismuth Bi 83 / 208.98	Polonium Po 84 / (209)	Astatine At 85 / (210)	Radon Rn 86 / (222)
Francium Fr 87 / (223)	Radium Ra 88 / (226)	Actinium **Ac 89 / (227)	Unnilquadium Unq 104 / (261)	Unnilpentium Unp 105 / (262)	Unnilhexium Unh 106 / (263)	Unnilseptium Uns 107 / (262)	Unniloctium Uno 108 / (265)	Unnilennium Une 109 / (266)									

*Lanthanide Series

Cerium Ce 58 / 140.11	Praseodym. Pr 59 / 140.91	Neodymium Nd 60 / 144.24	Promethium Pm 61 / (145)	Samarium Sm 62 / 150.36	Europium Eu 63 / 151.10	Gadolinium Gd 64 / 157.25	Terbium Tb 65 / 158.92	Dysprosium Dy 66 / 162.50	Holmium Ho 67 / 164.93	Erbium Er 68 / 167.26	Thulium Tm 69 / 168.93	Ytterbium Yb 70 / 173.04	Lutetium Lu 71 / 174.97

**Actinide Series

Thorium Th 90 / 232.04	Protactinium Pa 91 / 231.04	Uranium U 92 / 238.03	Neptunium Np 93 / (237)	Plutonium Pu 94 / (244)	Americium Am 95 / (243)	Curium Cm 96 / (247)	Berkelium Bk 97 / (247)	Californium Cf 98 / (251)	Einsteinium Es 99 / (252)	Fermium Fm 100 / (257)	Mendelevium Md 101 / (257)	Nobelium No 102 / (259)	Lawrencium Lr 103 / (260)

Table I Periodic Table of the Elements. [This is a modern version, included for reference. It is a far cry from the original version by Mendeleyev, Vernadsky's professor. Elements in *italics* have been created in laboratories of nuclear physics; all are highly radioactive. The names of these are interesting: After honoring Greek gods, places, and scientists, the namers apparently gave up with #104 and used a numbering system. *Ed*.]

chemical elements in the earth's crust; their forms of existence and classification

1 geochemistry classification of chemical elements

The first question that appeared in geochemistry was that of the number of bodies subject to its study: i.e., the number of different chemical elements and atoms that exist or possibly exist in the Earth's crust. At present we can only investigate this problem as far as it concerns the surface layer of the Earth. As for this area, the question can be answered definitely enough. In general, taking into account only the isotopes we know, and not the possibly existing ones, we can state more than 200 different compositions of atoms, corresponding to the 92 atomic numbers – N. Moselli's numbers – in the Periodic Table of D. I. Mendeleyev (table 1).

Within the limits of Mendeleyev's table, all the representatives of its 92 atomic numbers are apparently known; they are either isolated, or their existence on our planet is confirmed by exact data. But it is possible that in the world – on our planet – there are several elements that are not covered by this table elements lighter than hydrogen, for which Moselli's number is one, or heavier than uranium, which has atomic number 92. We do not understand as yet why the periodic system includes the number of elements we observe now (92), but it is inevitable for scientific thought to try to expand these limits, both by theoretical speculations and by experiment.[17]

---

17. Vernadsky was both prophetic and mistaken in his speculations about elements on Earth other than the 92 that make up the Periodic Table. He was prophetic in that several transuranic elements are now known (see modern version of the Periodic Table), having been produced in the laboratory, but he was mistaken in thinking that these might be part of Earth's geology. –*Ed.*

Up till now these efforts were unsuccessful. The attempt of V. Harkins to regard the neutron as a chemical element is sure to be proven unsuccessful too. But as long as it is possible to approach the solution of these problems experimentally, they should not be dropped. This concerns the possible existence of transuranic elements (93 and higher). We should assume that the number of elements (92) and the number of the known isotopes (219) is not final but only temporary, as has been stated empirically.

Geochemistry can study elements only in the thin surface layer of the Earth, which does not exceed 16 to 20 km and which comprises the upper part of the Earth's crust. I shall dwell upon this crust later, and we shall see that its total thickness reaches 60 to 100 km. The atmosphere is situated above it, but only its lower part, the troposphere, is chemically related to the Earth's crust: its thickness is 10–15 km. The height of the whole atmosphere (i.e., of the gases following the movement of our planet's body) is much more considerable, and it undoubtedly exceeds 700 km. The chemical elements of the Earth's crust and the troposphere are distributed in quite different ways. The difference in the quantity of the various elements contained in it is enormous. The quantity of oxygen – the most widespread element – exceeds the quantity of radium by hundreds of billions of atoms, but radium is not the rarest elementary substance of the Earth's crust.

Here I give a table (table 2) of the quantities or masses of the chemical elements contained in the Earth's crust (including the atmosphere), expressed both in weight percentage of the Earth's crust and in tons. The distribution of chemical elements in the Earth's crust is given in weight percentage and is categorized by orders of ten: The mass of the Earth's crust at a maximum thickness of 20 km is 3.25×10^{19} tons.

table 2   the abundance of chemical elements in the earth's crust as a percentage by weight

(the weight of the earth's crust at a maximum thickness of 20 km is 3.25 × 10^{19} t)

Categories	Mass in % (Weight T)	Mass in tons	Elements (Mass in %)
1	>10	>10^{18}	O (49.5 %), Si (25.7 %)
2	1–10	10^{17}–10^{18}	Al (7.5 %), Fe (4.7 %), Ca (3.4 %), Na (2.6 %), K 2.4 %), Mg (2.0 %), H (1.0 %)
3	10^{-1}–10^{0}	10^{16}–10^{17}	Ti (0.5 %), C (0.4 %), Mn (0.1 %), Cl (0.2 %), S (0.15 %), P (0.1 %)
4	10^{-2}–10^{-1}	10^{15}–10^{16}	N, Ba, B, V, Li, Ni, Sr, Cr, Zr, Br, Cu, F
5	10^{-3}–10^{-2}	10^{14}–10^{15}	Be, I, Sn, Co, Th, U, Zn, Pb, Mo, Rb, V, Ce
6	10^{-4}–10^{-3}	10^{13}–10^{14}	Ar, W, Cs, Bi, Cd, Hg, Hf, Nd, Sm, Gd, Yb, Pr
7	10^{-5}–10^{-4}	10^{12}–10^{13}	La, As, Nb, Sb, Ag, Se, Te, Tl, Eu, Er, Dy, Ta, Ho, Tu
8	10^{-8}–10^{-5}	10^{11}–10^{12}	Au, Pt, Ge, In, Os, Ir, Ga
9	10^{-7}–10^{-6}	10^{10}–10^{11}	He, Re, Ru, Rh, Pd
13	10^{-11}–10^{-10}	10^{5}–10^{6}	Ra

Comment: less than 10^{-11} % are Kr, Xe, Ne, Po, Pa, Ac and Rn.

This table was presented in its general features by the American scientist F. Clarke, who had studied these problems for more than forty years. I have introduced some corrections and changes and given it a different form. It is created on the basis of a huge number of exactly stated facts and many thousands of chemical analyses. The latest calculations of F. Clarke and W. Washington are based on 5508 complete chemical analyses of rocks that were done during the last 30 years. More than 100 years ago, in 1815, the English mineralogist W. Phillips was the first to make such calculations for 10 chemical elements. He returned to this task several times but his calculations, supported by D. Phillips and H. de La Beche, did not become part of science. Still, a small number of scientists, including Elie de Beaumont and A. Daubret, did not drop the task. Much later, in 1889, F. Clarke returned to this problem by systematically studying the principal chemical elements, and at the end of the nineteenth century, I. Focht tried to cover all the chemical elements in this way. Forty years is a sufficient amount of time to judge the correctness of this empirical generalization, and we must say that no significant changes have been made in F. Clarke's table since then.

Studying the table, we see that there is a correlation between the abundance of chemical elements in the Earth's crust and the composition of corresponding atoms. This correlation is very complicated and is not quite known to us. Prof. G. Oddo from Pavia had noticed long ago that chemical elements possessing even atomic numbers and containing nuclei of helium; that is, elements whose atomic mass is divisible by four, strongly prevail in the Earth's crust; they comprise 86.5% of its total mass. Later, similar investigations were made and deepened by Prof. W. Harkins in Chicago. Harkins proved that the same fact could be observed in meteorites, where the prevalence of elements with even atomic numbers is still more considerable. It reaches 92.22% for metallic meteorites, and 97.69 % for stony meteorites.

Meteorites are celestial bodies independent of the Earth, and maybe of the Solar System as well. Their chemical processes have a very indefinite and distant analogy with the processes of the Earth's crust. But the same regularity is observed in them – the same prevalence (even more pronounced) of elements with even atomic numbers, of atoms with even electric charges of nuclei.[18] Nevertheless, this very simple observation raises very important problems. It proves that the chemical composition of the thin surface film of our planet, which as far as we know does not at all correspond to the composition of the whole planet, is not accidental.

The chemical composition of the Earth's crust is connected with the definite structure of its atoms. Long ago, before Oddo's time, D. I. Mendeleyev had pointed out that the entire principal mass of the substance of the Earth's crust consists of light elements (not heavier than iron, #28). Apparently, if the even ordinal elements prevail, the even columns of the Mendeleyev table prevail too. The importance of these observations is evident, for they show that the chemical composition of the Earth's crust cannot be explained by geological reasons. But no important further conclusions have been drawn yet, and the main point is that the field of empirical observation was not expanded after all, in spite of numerous attempts. Hypotheses and extrapolations dominate here. Very often scientists point to the lesser stability of the nuclei of uneven atoms, but this too is a hypothesis.

This phenomenon may be connected with another one, for observations show that the surface parts of not only our planet, but also those of other celestial bodies – the Sun and stars – have a similar composition. This gives the impression that some regularities exist, which may be connected to the exchange of matter between all the outer envelopes of all the cosmic bodies. The existence of such incessant matter exchange is almost completely ignored now, although one can hardly doubt it.

---

18. These interesting observations are now explained by nuclear physics, specifically the synthesis of elements in star cores by nuclear reactions involving helium nuclei. –*Ed.*

The connection between atomic composition and the abundance of elements in the Earth's crust manifests itself distinctly in another phenomenon discovered not long ago by V. M. Goldschmidt. For the lithosphere – the Earth's solid crust – it is possible to calculate the volume occupied by different atoms. Proceeding from the numbers of Clarke and Washington for massive rock formations, and taking into account the fact that in crystalline silicates and alumsilicates the atoms are ionized so that an isotropic (spherical) field of application of their forces could be assumed for them, Goldschmidt calculated the volume occupied by atoms in the solid lithosphere (table 3).

table 3   percent by volume of atoms in the lithosphere

Atoms	Volume in %	Atoms	Volume in %
O	91.77	Al	0.76
K	2.14	Fe	0.68
Na	1.60	Mg	0.56
Ca	1.48	Ti	0.22
Si	0.80	--	--

Although the fields of atoms cannot have an ideal spherical form, the amendment will not change the principal conclusion about the distinct prevalence of oxygen and the rarity of silicon within the lithosphere's volume. The lithosphere consists mainly of oxygen atoms, and they are almost contiguous within it. A similar phenomenon is observed for the hydrosphere, which consists almost completely of oxygen by mass (88.89%).

The influence of the structure of atoms must manifest itself also in other properties of the Earth's crust, and must first of all be expressed in the scientific classification of natural bodies, in the "natural classification" as it was called in the eighteenth and nineteenth centuries. Any observational science is always based on such a classification, and geochemistry is one of these sciences. That is why we must begin an account of it with the classification of its objects – chemical elements – on the basis of studying the phenomena they create in the Earth's crust.

There is a premise necessary for such a classification: it should be constructed without any hypothesis in view. "A natural classification" is always strictly an empirical generalization, based without exception on scientifically proven facts. When the Periodic Table of chemical elements was being created, geochemical facts were not taken into consideration. That is why geochemical classification cannot be replaced by chemical classification. Geochemical classification should be based on the most general phenomena of the history of chemical elements in the Earth's crust – all particularities should be ignored.

The most general phenomena can be reduced to the following three characteristic features:

1. Presence or absence of chemical or radiochemical processes in the history of the given chemical element in the Earth's crust.
2. The character of these processes; their reversibility or irreversibility.
3. Presence or absence in the history of the chemical elements in the Earth's crust of their chemical compounds, or molecules consisting of several atoms.

As in all natural classifications, the limits between the groups may happen to be indistinct. Sometimes, for instance, one and the same chemical element can be placed into different groups. In this case, the history of the main part of the mass of the atoms or the most striking feature of their geochemical history will be crucial.

So, in the history of very radioactive elements (for example in the history of radium) we notice reversible chemical processes for its compounds, and irreversible radiochemical processes for its atoms. Radium will find its place in a group of elements for which the reversibility of the processes will be the most striking feature. I think that the general difficulties we shall come across here do not exceed those inherent in any natural classification, for classification inevitably leads to simplifying parts of nature that are indivisible and inseparable in essence.

At present it is impossible to classify only three elements from the viewpoint described above: the newly-discovered #43, and also the more familiar #85 and #87, whose masses have not been determined as yet. From this point of view, chemical elements can be subdivided into the following six geochemical groups (table 4). The percentages are related to the 92 elements of the Periodic System. The figures in subscript correspond to the atomic masses.

In all these groups, the difference between even and uneven numbers is evident. For groups 1, 4, and 5 it can be expressed quantitatively with sufficient precision (table 5), and for groups 1 and 5 this correctness is without doubt. For group 3, embracing the majority of the elements, it becomes noticeable only concerning widespread elements; that is, elements that make up a large portion of the total mass of matter.

table 4   chemical elements in geochemical groups

Geochemical Groups	Elements	Number	%
1. Rare Gases	He_2, Ne_{10}, Ar_{18}, Kr_{36}, Xe_{54}	5	5.44
2. Noble Metals	Ru_{44}, Rh_{45}, Pd_{46}, Os_{76}, Ir_{77}, Pt_{78}, Au_{79}	7	7.61
3. Cyclic Elements	H_1, (Be_4), B_5, C_6, N_7, O_8, F_9, Na_{11}, Mg_{12}, Al_{13}, Si_{14}, P_{15}, S_{16}, Cl_{17}, K_{19}, Ca_{20}, Ti_{22}, V_{23}, (Cr_{24}), Mn_{25}, Fe_{26}, Co_{27}, (Ni_{28}), Cu_{29}, Zn_{30}, (Ge_{32}), As_{33}, Se_{34}, Sr_{38}, (Zr_{40}), Mo_{42}, Ag_{47}, Cd_{48}, (Sn_{50}), (Sb_{51}), (Te_{52}), Ba_{56}, (Hf_{72}), (W_{74}), (Re_{75}), (Hg_{80}), (Tl_{81}), (Pb_{82}), (Bi_{83})	44	47.82
4. Dispersed Elements	Li_3, Sc_{21}, Ga_{31}, Br_{35}, Rb_{37}, Y_{39}, (Nb_{41}), In_{49}, I_{53}, Cs_{55}, Ta_{73}	11	11.95
5. Elements of High Radioactivity	Po_{84}, Rn_{86}, Ra_{88}, Ac_{89}, Th_{90}, Pa_{91}, U_{92}	7	7.61
6. Elements of Rare Earth	La_{57}, Ce_{58}, Pr_{59}, Nd_{60}, Pm_{61}, Sm_{62}, Eu_{63}, Gd_{64}, Tb_{65}, Dy_{66}, Ho_{67}, Er_{68}, Tu_{69}, Yb_{70}, Lu_{71}	15	16.30

table 5

Groups	Even Elements (%)	Odd Elements (%)	Distribution of Mass (%)		Mass of Earth's Crust
			Even Elements	Odd Elements	
1	100	0	100	0	$n \times 10^{-6}$
3	56.82	43.18	85.8	14.2	>99.7
4	0	100	0	100	$>3 \times 10^{-2}$

For the other three groups the data are less precise. But F. Clarke's table, which was presented long before the appearance of our ideas about atomic numbers and the positive charges of nuclei (and quite independently from them) shows that the elements of these groups, which are comparatively widespread, correspond to the even atomic numbers (table 6). So the prevalence of the mass of chemical elements with even atomic numbers is quite evident in five groups of natural classification; only group 4 does not include elements with even numbers.

table 6

Groups	Even Elements (%)	Odd Elements (%)	Distribution of Mass (%)		Mass % of Earth's Crust
			Even Elements	Odd Elements	
2	55.55	44.45	?	?	3×10^{-5}
5	71.4	28.6	>99.9	$<10^{-1}$	$>5 \times 10^{-3}$
6	50.0	50.0	>99.0	<1.0	$>1 \times 10^{-19}$

The first group – that of rare gases – includes elements that take no part in the main terrestrial chemical processes, and that make up compounds with other elements only in exceptional cases. These atoms are preserved practically unchanged throughout geological time. A closer study of their history makes us discard the early ideas of C. Moureux, who suggested that they are absolutely inert in geological history, and that in them we observe the remains of the cosmic history of our planet. The quantitative intensity of their chemical manifestations in the thermodynamic field of our planet is so different from other compounds, so relatively small, that their actual difference from other terrestrial elements cannot arouse any doubt. However, their geochemical significance is enormous, and their role in the worlds beyond the Solar System must be great, too.

One of them – helium – is very widespread in the substance of celestial bodies, and apparently it plays a significant role there that has not yet been discovered. Its quantity in the Earth's crust is changeable and seems to be increasing, as it is continuously appearing there due to decomposition of the nuclei of uranium, ionium, radium, radon, RaA, RaC, RaCl, polonium, thorium, radiothorium ThX, thorone, ThA, ThC, protactinium, radioactinium, AcX, actinone, AcA, AcC, AcCl, samarium, and possibly beryllium. It is expected that the process does not stop here, and that there are other elements that secrete alpha particles (just as helium atoms carrying two charges while decomposing, eventually lose their charges and transform to ordinary helium gas).

But there are cases in which the rare gases, called this way by chemists because of the difficulty of creating their chemical compounds under the conditions of our laboratories, do give compounds. These compounds, in the form of water solutions and hydrates as was recently shown by V. G. Chlopin, must play an important role in the structure of the biosphere. Finally, they may include also radon, a rare gas from group five, which is a carrier of great active energy in its different isotopes. In general, the role of the rare gases in the structure of our planet is much greater than their relatively small quantity; and this role is just beginning to reveal itself to us.

The second group – that of inert elements of noble metals in the Earth's crust – includes the two last columns of D. I. Mendeleyev's Periodic Table of elements; gold can be included here too. These elements give an almost infinite number of compounds in our laboratories and this is their difference from the rare gases. But their compounds are almost absent in the Earth's crust. The minerals corresponding to them, mainly alloys, existing because of a complicated pneumolythic and magmatic process, or (for gold) because of abyssal hydrothermal processes in thermodynamic conditions that are distinctly different from those of the biosphere, change very little or not at all in the course of geological time. This stability, as well as that of the rare gases, is not complete. For some small part of their terrestrial mass, very slow chemical reactions must exist that change them, and these reactions are not well studied.

For instance, in the biosphere, oxygen compounds of palladium emerge. For palladium and for the nuggets of platinum and gold, there are numerous phenomena of weathering connected with the re-crystallization and change of the chemical composition of the alloy. For gold, their phenomena are connected with decomposition of telluride compounds. But these slow and local chemical reactions do not change the general character of the group – its terrestrial chemical inertness. It is furthermore characteristic of the whole group that these elements are only slightly affected by the aquatic structure of the Earth. They find themselves in a dispersed state in water solutions or connected with phenomena of sorption.

The third group of cyclical or organogenic elements is the largest in mass. It includes the greatest number of chemical elements and makes up almost the entire Earth's crust. It is characterized by numerous reversible chemical processes. The geochemical history of these elements may be expressed by cycles. Each element gives compounds characteristic of a certain geosphere; these compounds are constantly being renewed. After more or less long and complicated changes, an element returns to its initial compound and begins a new cycle. This character of terrestrial chemical reactions was noticed for oxygen in the second half of the eighteenth century; the great scientists of that time, who had discovered the terrestrial gases and their properties, foresaw these characteristic chemical cycles.

I think that Dr. J. Pringle, the President of the Royal Society in London, was the first to express these notions in 1773, in his speech about J. Priestley. He defined the general features of the great equilibrium of vegetative green chlorophyll matter, together with animal matter, in relation to free oxygen and carbon dioxide. In 1842, two French scientists – J. B. Dumas and J. Boussingault – gave a clear picture of these cycles, and in the 1850s C. G. H. Bischof, J. Liebich and K. Moor transferred these notions to the rest of the matter of the Earth's crust. Since then, science has collected a great quantity of empirical facts confirming these generalizations. These facts were not coordinated though, and are in a state of almost complete chaos. The importance of living matter for these cycles is being confirmed. This importance is observed not only for organogenic elements, such as C, O, H, N, P, and S, but also for metals such as Fe, Cu, Si, V, Mn, etc., and for all the chemical elements of this group, as we shall see.

The elements of this group are part of cycles that are characterized by chemical compounds, molecules, or crystals. These cycles are reversible only for the main part of the atoms involved, some of the elements inevitably and continually leave the cycle. This is natural; that is, the cyclical process is not completely reversible. Among such ways of leaving the cycle, the most significant dispersal of an element is its exit in the form of free atoms. In this way the element may leave the cycle forever. Still, it is clear that even if future discoveries more or less alter our present-day ideas, they will not deny the main empirical generalization regarding the prevalent significance of chemical compounds and reversible cycles in the history of the main mass of the Earth's crust. The cyclic elements are included and play an important role in the aquatic apparatus of the Earth's crust: they are included in water solutions (ions), and make up minerals formed by water. Only zirconium and hafnium seem to stand aside in this respect. Zr and Hf do not enter living matter, and germanium has not yet been found in it either, but judged by its aquatic history, it surely will be.

In the next group, that of dispersed elements, free atoms prevail. They cover a small part of matter, and they also have their cycles, which renew constantly. Not always though are they expressed by chemical compounds, by molecules; their compounds decompose more or less completely in one area of these cycles and renew under different thermodynamic conditions in another area. All the dispersed elements are characterized by the absence or rareness of chemical compounds, not only in certain areas of the Earth's crust, but in the Earth's crust as a whole.

There are two cases that are distinctly different from each other. Some of the elements, such as Li, Sc, Rb, Y, Cs, Nb, Ta and maybe In, form chemical compounds only in deep zones of the Earth's crust. Their minerals are located in the surface area in the biosphere, but the new compounds of these elements – new minerals – are not formed here; the elements do not form vadose[19] minerals. Instead, the elements are dispersed throughout the surrounding substance as "traces," as analysts say, and have seemingly nothing to do with the mountain rocks they are found in.

The second case is that of iodine and bromine. They enter compounds with other elements only in the biosphere, which means that all their minerals are of vadose nature. If we try to reconstruct their history and find out their origin, we shall make sure that the sources of iodine and bromine are water solutions, and that living matter has extracted and concentrated them from those very solutions. In the depths of the crust we find iodine and bromine only dispersed as traces in minerals or in rocks – both metamorphic and plutonic – without any apparent relation to their chemical composition. Our knowledge is not sufficient to fully discover the history of gallium, but apparently it belongs to the second group as well. At the present time, its compounds are not known. The maximum content of gallium in a mineral – germanite – does not exceed 7×10^{-10} % of metal, and in micas its content reaches the same order.

All these are minerals of the deep regions of the Earth's crust. Hence, the cyclic processes corresponding to these elements are specific; the elements give chemical compounds and free atoms in turn. But the majority of them do not enter compounds at all. They are constantly dispersed everywhere in the matter of our surroundings, apparently in the state of free atoms. They appear to be in a state close to that of rare gases, outside chemical reactions in the parts of the planet acces-

---

19. A dictionary definition of the *vadose zone* is: The region of aeration above the water table. This zone may be absent to several hundred feet thick, depending on the environment, the type of Earth material present, etc. As here, Vernadsky expands this definition to include any zone that might be suitable for life. *–Ed.*

sible to our investigation. The fact that all these elements belong to one and the same group, to that of atoms with uneven atomic numbers, evidently shows that the structure of these atoms has peculiar characteristics connected with this way of spreading.

This phenomenon deserves much more attention than is usually paid to it. Such a state of chemical elements can bring about processes of great cosmic importance. If it is the common property of elements with uneven atomic numbers, it can explain the prevalence of their antipodes – even elements – in the Earth's crust and meteorites. All the uneven elements, except for Sc, Nb, and Ta, take part in the aquatic regime of the planet by being there in a dispersed state. Some of them, such as Li, Br, and I, are concentrated by living matter; Sc, Ga, Y, Nb, In, and Ta are concentrated by organisms that have not yet been studied.

The fifth group of elements includes very radioactive elements: the families of uranium, actinouranium and thorium. Here the incomplete reversibility of processes is quite evident. In general, uranium and thorium make up compounds included in reversible cycles, the closed cycles, which are analogous to the cyclic processes of the cyclic elements. But part of their atoms is lost in the course of the cyclic processes and does not return; it gets decomposed, changes and gives birth to other elements, two of which, helium and lead isotopes, belong to the groups of rare gases and cyclic elements, which are quite different chemical groups.

Now it is becoming clear that radioactive decomposition is characteristic not only of heavy atoms, but of light atoms as well. In 1907, Campbell discovered two radioactive elements with beta-radiation: potassium (from the group of cyclic elements) and rubidium (from the group of dispersed elements). In the case of rubidium, atoms of strontium must appear (belonging to a different geochemical group), and in the case of potassium, atoms of calcium (belonging to the same group) and of argon. Twenty-five years later, another period of discoveries began, in which von Hevesy and Pahl discovered the radioactivity of samarium, belonging to the group of rare Earths; it transforms to neodim through alpha-radiation. We seem to be on the verge of great discoveries.

It is quite probable that frailty is a property of all elements. Even if these probabilities become scientifically proven facts, it will not affect the specific position of the group of radioactive elements in the system of classification. Decomposition of elements in this group is quantitatively incomparable with its possible manifestation in all other elements. Weak radioactive elements, in their geochemical manifestation, can be united into one group with strong radioactive ones as little as ferromagnetic elements can be united with the usual paramagnetic ones in case of magnetic properties.

The last group – that of rare Earths – must here and in the Periodic Table of chemical elements be presented as a special group. I think it consists of 15 elements that correspond to atomic numbers 57 to 71 without a break. Scandium and yttrium are sometimes included into this group although they do not really belong there. As we have seen, they belong to the group of dispersed elements. Concerning scandium, this seems certain to me from the chemical point of view. As for yttrium, some chemists, for instance R. Vogel, have come to the conclusion that it should be separated from the rare Earths for purely chemical reasons.

From the geochemical point of view, the main characteristic feature of these elements is the complete absence of their vadose compounds (compounds that have appeared in the biosphere). But their history in the biosphere is not quite clear as yet. It is evident that some of them get dispersed in it: for instance, gadolinium, samarium, europium, and neodymium. They, as well as cerium and lanthanum, enter living matter where their history is unknown. But at the same time, their principal minerals such as monacytes, xenotymes, and orthites, which appeared in magmas or pegmatite veins under conditions of high temperature and high pressure, are very stable in the thermodynamic field of the biosphere, which is quite different from those original conditions. It is possible that the majority of their atoms stay inert there and do not migrate.

There are indications of genetic correlations between the elements of these groups, but these indications are, until now, beyond the realm of facts. But one essential fact arouses no doubt: All the elements of this group, "the chemical nebula" as it was called by Crookes, usually remain together in one body under diverse terrestrial conditions since they do not react with the majority of terrestrial chemical elements. The question is being solved now. The observations of von Hevesy seem to indicate the genetic radioactive connection between samarium and neodymium, and further investigations will reveal more. But even if radioactivity – the weak type – is proven, it will not interfere with the isolation of this geochemical group. The elements of this group do furthermore not comprise any noticeable part of the aquatic structure of the Earth's crust. Minerals coming from water solutions are not known.

The quantities of matter concentrated in each of the six geochemical groups of elements are very different (table 7).

table 7 the masses of geochemical groups of elements in the Earth's crust.

Geochemical groups	Weight in tons
Rare Gases	10^{14} t
Noble Metals	10^{12} t
Cyclic Elements	10^{18}–10^{19} t, close to 2×10^{19} t
Dispersed Elements	10^{16} t
Elements of High Radioactivity	10^{15} t
Elements of Rare Earths	10^{16} t

Of course this table can be regarded as a first approximation to reality, but the order of the phenomena is expressed rather exactly. The cyclic elements comprise more than 99.7%, almost the whole mass of the Earth's crust. But the remainder of 0.3% is not an insignificant quantity. It makes up quadrillions of metric tons. It includes, for example, the radioactive elements, whose great significance in the mechanism of the biosphere will become clear further on. It refers to matter in a chemically active state that possesses free (atomic) energy, and it therefore performs an enormous amount of chemical work in the Earth's crust. The quantity of this matter is measured by a number of the order of 10^{15} t. This number is close to the mass of another kind of "active" matter of the Earth; that is, living matter (living organisms), which is as deeply implanted into the mechanism of geochemical processes. In fact, the small fractions of the Earth's crust that correspond to them bring about all the grand geochemical (and apparently a lot of geological) processes of our planet.

2 forms of existence of chemical elements

The history of chemical elements in the Earth's crust can always be reduced to their various movements, or shifts, that in geochemistry we shall call *migrations*. The movements of atoms making up compounds, their transmissions in liquids, gases and solid bodies, and in the processes of breathing, nutrition, the metabolism of organisms, etc., are all migrations. These migrations within the Earth's crust create large systems of various chemical equilibria.

In geochemistry, the principal task is the study of equilibrium systems resulting from the elements' migrations. These systems can always be expressed in terms of mechanics, and in the form of dynamic and static systems, those of atomic equilibria. The laws of equilibrium, of homogeneous and non-homogeneous systems of any kind of bodies, embrace the whole of geochemistry. The profound synthesis of these laws was made at the end of the last century by the American scientist J. W. Gibbs, and was deepened by the investigations of G. Duham, H. le Chatelier, G. V. Backius-Rosenbum, K. Brown, G. H. Tammann, and others.

In the history of chemical elements of the Earth's crust, we can separate several different groups of equilibrium systems, which can contain the elements for an indefinite amount of time – "eternally" on the scale of geological time. These groups of equilibrium systems are more or less independent and a chemical element is subject to different physical and chemical regularities in each group. The study of geochemical problems can be reduced to the study of the history of every chemical element in the conditions of each of these groups, and to the mutual correlation between the histories traced in such a way, because a characteristic feature of the terrestrial history of the chemical elements is the incessant migration of elements from one equilibrium group to another throughout geological time.

I will refer to these different groups of equilibrium systems as different forms of existence of chemical elements. I cannot give a detailed account of the forms of existence of chemical elements here. The only thing I will say is that these forms must be quite numerous, but that not many of them can be observed on the Earth, to say nothing of the Earth's crust. If we go beyond the limits of the Earth's crust, and moreover, beyond the limits of our planet, we will come across alien forms of existence of chemical elements that are unknown to us here. Among them are the "gases" of solar corona electrons, the states of a comet or nebula substance, and heavy gases of stars such as star B of Sirius. The forms of existence have been determined in a purely empirical way, and each of them has turned out to contain atoms in specific states. In fact, they are fields of different states of atomic systems.

In the Earth's crust, we distinguish four different forms of existence for chemical elements. First, the following three:

1. Molecules and their compounds in minerals, rocks, liquids and gaseous terrestrial masses.
2. The existence of chemical elements in living beings; the autonomous manifestations of living matter.
3. The existence of elements in silicoaluminum magmas; complex, ever-changing systems, more or less viscous, which have a high temperature and a high pressure and which are supersaturated with gases.

We can clearly imagine neither chemical processes nor states of atoms in these media, which exist in a thermodynamic field of phenomena that is alien to us. But there is one more, the fourth form of existence, which is usually not distinguished and not taken into consideration – that of the dispersal of chemical elements. In migrations of elements this form plays a very important role. As we have seen, it is typical of a certain geochemical group of elements. While studying the geochemical history of elements, none of these forms of existence can be ignored. We must consistently study the fate of each element in all of them and pay special attention to the migrations of elements from one form to another, which are not at all accidental. As the significance of the elements' dispersal is usually not realized from this point of view, I would like to make it clear by dwelling upon the history of two chemical elements that belong to the group of dispersed elements: iodine and bromine.

3   geochemistry of iodine and bromine

From the everyday experience of our laboratories, we know that iodine and bromine can make up thousands of compounds with other simple bodies. Many of these compounds are very stable in the thermodynamic conditions of the Earth's crust, but they do not appear there. We can find only 13 minerals containing iodine, and 3 to 4 minerals containing bromine. The quantity of bromine in the Earth's crust is no less than 10^{16} t, and the quantity of iodine is about 10^{15} t. Iodine and bromine are much more widely spread, by thousands of times more, than antimony, selenium or silver, but the number of minerals of which the latter are part exceeds 100 for each of them, while for bromine and iodine it is not larger than 17. There are five dubious minerals with iodine, and not a single mineral containing either bromine or iodine was found in large quantities.

Hundreds of thousands of tons of iodine are contained in iodic-acid calcium, perhaps also in iodic-acid sodium, and in very little investigated minerals of which lautarite is the most well studied. These iodic-acid minerals are dispersed in the saltpeter and gypsum deposits of South America. Iodide and bromic compounds of silver exist in smaller quantities and seem to be more stable in the lower areas of the biosphere below the oxygen surface. All other minerals of bromine and iodine are mineralogical rarities; sometimes they are found in kilograms or even smaller quantities. The quantity of iodine and bromine in minerals probably does not exceed a maximum of several million tons. The quantity of iodine in isomorphic admixtures seems to be even less. On the whole, the quantity of iodine in minerals is small compared to its mass existing in the Earth's crust. All minerals of iodine and bromine are vadose (i.e., they appear and exist only in the upper layer of the crust, in the biosphere). Their volcanic forms should also be considered vadose, since they cannot exist in deeper layers of the Earth's crust in a solid state. There is not a single mineral of iodine or bromine that has appeared in deeper metamorphic or magmatic parts of the crust.

Large quantities of iodine and bromine are contained in all organisms of living matter, as was discovered by W. Courtois. Many organisms seize it rapaciously. For instance, in the opinion of A. Gautier, all the iodine of the biosphere – of the surface layers of the ocean to a depth of 800 m, the soil and the atmosphere – exists only in that state; that is, gathered in living matter. A considerable quantity of iodine (and of bromine) is concentrated in aquatic terrestrial solutions, where it exists as ions I and IO^{-3}, and as free iodine. Organisms draw and concentrate their iodine (and bromine) from solutions, such as the deeper layers of the sea, lakes and saline swamps, salty springs, surface waters, and all fresh waters. Iodine and bromine penetrate into all the waters of the Earth's crust, not only into those of the biosphere. From time to time they concentrate in mineral springs or in stratum waters. The source of the iodine in water solutions is to a considerable extent its biogenic migration, which apparently is a general phenomenon. Iodine may reach the order of 10^{-2} % when it exists in these waters in the forms of ions (I^-), and free iodine.

Millions and millions of tons of free iodine do not enter either organisms, metals or natural waters; they are dispersed in rocks. According to A. Gautier, all the erupted and metamorphic rocks and the minerals contained in them, to say nothing of sedimentary rocks, contain iodine in the form of traces (in quantities of 1.7×10^{-5}–1.25×10^{-4} %). According to the new analyses of G. Fellenberg and his colleagues, these quantities are smaller (1.9×10^{-5}–8.1×10^{-5} %). The dispersal of iodine is extensive, and in this respect it resembles the existence of radioactive elements to which we are already acquainted. We may think of iodine as a model of radium; it can be found in all minerals without exception, and it is present in different, sometimes relatively large quantities. The numerous measurements of Fellenberg indicate a range from 3.8×10^{-3} (bornite) to 5×10^{-6} % (calcite).

The dispersal in the form of such "traces" is the most typical and usual form of existence of iodine in the Earth's crust. We cannot state any correlation between its quantity and that of other elements of the rocks and minerals in which it is found, as if the atoms, or maybe the ions of iodine, are dispersed throughout the terrestrial substance under the influence of physical instead of chemical forces, and maybe of interatomic ones. Nevertheless, it is quite probable that iodine and bromine exist in capillary water permeating terrestrial solid matter as a weak solution. The existence of iodine in organisms, waters, and solid bodies is closely related to its content in the terrestrial atmosphere, from which it can penetrate back into water and organisms through atmospheric fall out. Our knowledge of bromine's spreading is less complete, but it is clearly generally similar to the history of iodine.

Here we see a closed cycle of a new type. Iodine and bromine in a dispersed state become part of the substance of the Earth's surface. Their atoms or free ions are captured by living organisms and concentrated in the compounds they make up, which contain up to 8.5% of iodine and sometimes more, as for example the bodies of sea sponges.[20] Apparently, part of the dispersed atoms of iodine is also seized by chemical reactions of the surface, and makes up vadose minerals. It is quite possible though, that these chemical reactions can take place only in direct or indirect relation to living matter, or that they are observed only under conditions that are favorable for accumulating organic matter, the product of life. In the course of time, iodide and bromic products of organisms, as well as vadose organic minerals, which are always related to life, decompose. Iodine and bromine thereby return to the state of atoms and ions in order to begin the same cyclic process again.

---

20. It is interesting that Vernadsky does not mention the hormone, *thyroxine*, which comes in two forms containing either three or four atoms of iodine. Thyroxine controls the rate of metabolism (i.e., oxygen consumption) and is thus highly critical to human health. The presence of iodine in thyroid glands was demonstrated in 1896 by Bauman, and thyroxine was synthesized in 1927 by C. R. Harington and G. Barger. Too little iodine in the human diet leads to abnormal growth of the thyroid gland: a large swelling in the throat called goiter. Goiter was very common in Vernadsky's time, especially in land-locked countries where iodine-rich seafood was seldom consumed. Introduction of iodized salt has made goiter a rare occurrence in our time. –Ed.

Two phenomena are typical in this cyclic process: the influence of life and the weak concentration of iodine. It takes a long time for the cyclical process to be completed. Lately a new factor has appeared, which is humans. Throughout geological time, the iodine collected by life remained intact in coal with a weak concentration of about 6×10^{-4} %. Now, humans burn coal and, in such a way, introduce many thousands of tons of iodine into the atmosphere, which, after a million years returns to living matter. Another process of this kind concerns the iron ores that are always rich in iodine; they appear in the biosphere through an organogenic method or with the participation of life.

Such a cyclic process, covering a geologically long period, involves a small number of iodine atoms at a time. The main mass of iodine atoms is in a state of complete dispersal. This example shows that studying only minerals and rocks is far from giving us a complete picture of the existence of chemical elements in the Earth's crust. It should also include the most widespread elements constituting the Earth's crust – the cyclic elements. Here, not dispersal, but living matter, living organisms, plays a conspicuous part. The history of chemical elements cannot be understood without it.

4   living organisms in the earth's crust

From the geochemical point of view, living organisms are not an accidental phenomenon in the chemical organization of the Earth's crust; they make up its most essential and integrated part. They are inseparably connected with the inert matter of the Earth's crust such as rocks and minerals. In the majority of their papers, biologists who study living organisms ignore the inseparable and functional connection existing between a living organism and its surroundings. Although realizing clearly the organization of the organism, they fail to realize the organization of the surroundings in which the organism lives (i.e., the biosphere.) They see these surroundings as inert and independent of the organism, "cosmic," as Claude Bernard has well put it.

So by studying an organism they do not study a natural body, but an ideal product of their thought. Often it is a convenient, even necessary method of scientific work that is very widely adopted in the natural sciences. According to this method, complex natural phenomena are replaced by simplified models, and empirical conclusions and facts are idealized and deviated from. A material triangle is not the triangle of geometry, the "atmosphere" of physics is not the troposphere surrounding us, and an animal or plant of a biologist is not a real living body, not a natural organism. This must always be understood, and sooner or later the moment comes when it is necessary to crucially change the principal ideas. This moment is about to arrive for the biological sciences.

The great biologists of the past realized the inseparable connection of an organism with its surroundings. In the late eighteenth century, F. Vique d'Azir brilliantly expressed these ideas in the lectures he delivered in Paris, in which he tried to introduce a scientific and logical definition of life. This was one of the numerous definitions of life, one of the attempts to solve the problem (more a logical than a scientific one), which had drawn the attention of scientists and philosophers for many generations. In his influential report on the state of sciences in post-revolution France, presented to Napoleon in 1808, Cuvier expressed the same thoughts as Vique d'Azir with a peerless clarity and precision. He deepened these thoughts in his other papers. He wrote, "So life is a more or less fast, more or less complicated whirlpool, which always captures molecules possessing certain qualities, and which has a constant direction. But it is always penetrated by, and always deserted by individual molecules, so the form of a living body is more significant for it than its substance. As long as this motion exists, the body within which it occurs is alive, it lives. As soon as the motion comes to a complete standstill, the body dies." One of the main thoughts expressed here, that of the greater importance of the form of a living body than of its substance, was the principal idea of biology throughout the whole nineteenth century. It was accepted, but everything else significant for Cuvier was discarded. In fact, in the nineteenth century, not only the substance, the molecules, were put aside, but also the influence of an organism on the environment, i.e., the motions of the molecules of the environment that, according to Cuvier, were essential for life.

In the late nineteenth and the early twentieth centuries, the notions of the relations between life and environment, or the understanding of Cuvier's formula, became more profound. The roots of these changes are to be found in geological and biological research. Geologically, they brought forth the discovery of the organized character of the biosphere adjusted to life and regenerated by life. Its particular manifestations are the biogeochemical processes studied in this book. The whirlpool of atoms entering and leaving a living organism is determined by a definite organization of life's environment, by a geologically definite mechanism of the planet – the biosphere.

Biologists come to the same conclusion proceeding from the living organism. The form becomes clear only when both parts of Cuvier's whirlpool are taken into consideration: that in the environment and that in the morph [form] the organism. The outstanding and original French zoologist, F. Houssait, was quite right to point out "schematization"; the complete discrepancy between the biologists' "living organism" and a real living organism. The real organism is inseparably connected with the environment and can be separated from it only theoretically.

One cannot study and understand an organism, comprehend its form and vital activities, without studying and knowing its environment. A younger contemporary of Houssait, the English physiologist D. Haldan, insisted on an even closer connection between the environment and the organism's functions. Another physiologist, the American L. Henderson, put these concepts into a distinct and more profound form; he connected them into a single unity with geological processes. But all these scientific studies of biologists could not channel scientific biological work into a different direction.

The grand manifestations of living organisms that are evidently connected with the environment – their respiration and nutrition – continued to be studied and have been studied while disregarding their influence on the environment from which the organisms receive chemical elements, and to which they return them by means of these processes. A living organism of a biologist in the "cosmic" environment is, in the greatest majority of cases, different in its scientific scope from the real body of empirical knowledge – a living organism of the biosphere.

Whole fields of biological problems have remained outside the realm of biology. But separate thinkers among the biologists have long been trying to go deeper, to approach the general substratum of inert and living matter while handling biological problems. In the archives of science we find profound ideas of this kind, which should draw the attention of our time as well. These ideas are revivified now. The old scientists of the late eighteenth century were less limited by patterns and habits of mind than their descendants. Before the new chemistry was finally formed, the idea of a universal cosmos had dominated, and consequently a search for a universal power ruling the world. In all the phenomena of our life, such as an apple falling from a tree, and in the greatest cosmic manifestations, such as the movement of celestial bodies, hence in the whole system of nature, one and the same universal power was seen: gravity. I. Newton, who had discovered the law of gravity, and who ruled supreme in scientific understanding of the cosmos in the late eighteenth and in the nineteenth century, had not been the author of the idea of "universal gravity," but he also tried to transfer his laws to new fields such as chemistry, where it is not they that dominate as we will see.

In the eighteenth and nineteenth centuries, the manifestations of "universal gravity" were looked for everywhere. This led to the discovery of new laws and to the clarification of complicated and tangled phenomena, but at the same time the scientists always came to the conclusion that the newly discovered forces were different from "universal gravity." In the eighteenth century, Coulomb proved that the laws of attraction and repulsion of electrified bodies were similar to those of gravity, but that these laws are similar only in their outward appearance. Proceeding from gravity, Laplace came to the theory of special capillary forces. Attempts to find manifestations of "universal gravity" in the phenomena of chemistry and in chemical affinity brought forth the discovery of new laws and fruitful generalizations that had nothing to do with Newton's attraction of gravity.

In 1782, the St. Petersburg Academy of Sciences put forward a problem for a competition in the realm of biology, generated by the same trend of thought. The question was whether there were any relations between Newton's gravity and the force acting in the processes of nutrition and respiration of living matter. Which force makes it possible for living organisms to extract from their environment all the substance necessary for them to live and grow? Caspar Wolff, one of the most outstanding members of our Academy, a great investigator of life and one of the creators of embryology, initiated this question and published a memoir after the competition, in which he proved what seems obvious to us – that the forces of nutrition and respiration are quite different from Newton's attraction of gravity. But the thing that can intrigue us in this forgotten episode of the past is the question itself. In fact, this question is an attempt to scientifically embrace the reflection in the environment, in the biosphere, of the countless minute phenomena of respiration and nutrition of living beings. Respiration and nutrition are considered not only as phenomena of the organism but also as planetary phenomena.

In the late eighteenth and early nineteenth century, another scientist, the well-known Polish medical man J. Sniadecky, returned to the same ideas. He compared Newton's attraction of gravity with the "attraction" of matter – the respiration and nutrition of living beings. He expressed the idea that the intensity of the force inducing these processes in an organism grew in inverse proportion to the mass of the organism, while Newton's gravity acts in direct proportion to the mass. Small living beings unseen by the eye produce the most significant effects. This trend in biological research soon died out completely, but the complex of ideas that had induced it has lately reappeared in geochemistry because the influence of living beings in the history of chemical elements of the Earth's crust is exerted mainly by their nutrition and respiration.

In geochemistry, organisms manifest themselves and can be studied only from the point of view of the general effect created by these physiological phenomena, the complex of which makes up a planetary phenomenon. But these ideas are even more profoundly connected with our thought. They are most closely related to a biogeochemical study of life, as in the research of Wolff, Sniadecky, and earlier in that of Buffon. The universal gravity by Wolff, and the atoms in life's biogeochemical embrace by Sniadecky, are attempts to connect the phenomena of life with the main elements that manifest themselves in the cosmos.

5 the history of free oxygen

The planetary significance of the phenomena of life, namely of respiration, can be well understood if we consider the history of free oxygen in the Earth's crust – one of the innumerable chemical bodies introduced into the biosphere by living matter.

Free oxygen as molecular O_2 in gaseous form, and especially in water solutions, plays quite an exceptional role in all the chemical reactions of the Earth's surface. We can even say that its presence changes the course of these reactions. The quantity of perpetually existing O_2 molecules in the Earth's crust is enormous. It can be determined with sufficient precision. In the atmosphere (the troposphere and the lower stratosphere), the weight of free oxygen or O_2 molecules amounts (according to S. Arrhenius) to a minimum of 1.2×10^{15} t and a maximum of 2.1×10^{15} t. This mass exceeds that of many chemical elements in the Earth's crust by hundreds of thousands of times.

The atmosphere does not contain all the free oxygen; a considerable part is dissolved in waters and mainly in salty water, which makes up the world's ocean. This amount hardly exceeds 1.5×10^{13} t. Free oxygen is also dissolved in the fresh water on land and dissolved or occluded in snow and ice. But this quantity is much smaller than that of the marine part of the hydrosphere, because according to W. Halbfass, the volume of fresh water makes up only 3.6×10^{-10} % of the volume of ocean water, even when it includes snow and ice, which present the dominating part of the weight of terrestrial water. So, according to W. Halbfass, the volume of ice corresponds to $3.5-4 \times 10^6$ km^3, the volume of ocean water to 1.3×10^9 km^3 (O. Kruemmel), and the volume of the lake, swamp, river, and surface waters maximumly to 7.5×10^5 km^3. Furthermore, the whole amount of free oxygen included into sedimentary rocks slightly exceeds 1.5×10^5 t and comprises approximately $1/10^5$ of all the oxygen of the Earth's surface.

We know that free oxygen exists only on the Earth's surface. The water of deep springs does not contain it, as was proven in the late eighteenth century by the English physician D. Pearson (1751–1828). The gases of volcanic and metamorphic rocks are almost devoid of it too. The quantity of free oxygen in the biosphere is undoubtedly one of the most precisely estimated physical constants of our planet. It determines the geochemical work of living organisms and allows us to understand the significance of free oxygen in the history of chemical elements. Free oxygen is the most powerful agent among all known chemical bodies of the Earth's crust; it changes, or oxidizes, a great quantity of chemical compounds, it is in constant motion, and it constantly forms compounds. We know thousands of chemical reactions by which it is captured, and during which it enters the compounds. The most significant ones among these compounds are the oxidized forms of metalloids such as sulfur and carbon (including the compounds of organisms) and the compounds of metals such as iron or manganese.

The history of all cyclic elements of the Earth's crust is determined by their relation to free oxygen. Recent investigations have even pointed out its major influence in volcanic phenomena. The atmospheric oxygen captured by burning lava yields oxidized products (such as waters, sulfur oxides, etc), and the heat liberated by these reactions plays a most significant role in the thermal effects of lavas. The temperature of lava rising from the entrails of the crust, which has not yet contacted the oxygen of the air, is often hundreds of degrees lower.

In spite of the significance of these reactions for numerous terrestrial processes of this kind, the overall amount of free oxygen on the planet seems constant or almost constant. Evidently, some reverse processes must exist, which liberate free oxygen into the surroundings. We know only one reaction of this kind in the biosphere, if we consider large-scale reactions. This reaction is biochemical; it is the release of oxygen by chlorophyll plastids of terrestrial organisms. This reaction was discovered in the late eighteenth century by J. Priestley, subsequently deepened by the works of outstanding scientists of that time, and presented in all its significance, its general character and its main features, by the scientist T. de Saussure of Geneva at the beginning of the last century.

This reaction of forming free oxygen in the Earth's crust is undoubtedly not the only one, but as far as we can judge, it is the only one releasing considerable masses of free oxygen to the structure of the atmosphere that envelopes our planet. The excretion of free oxygen outside the influence of life is proved to be, or most probably, due to the processes of radioactive dissociation, decomposition of gases by ultraviolet radiations, and metamorphism. Oxygen must be isolated in the depths of the Earth's crust, since the compounds rich in oxygen appearing on the surface, such as sulfates and bodies containing ferric oxides, turn into compounds that are poorer in oxygen, or which do not contain it at all, upon reaching the deep layers of the crust. But this free oxygen must immediately enter compounds, for we fail to see its manifestations anywhere.

Even if from time to time and at some places free oxygen rises from the depths of the Earth's crust, its mass in the biosphere is insignificant in comparison to the amount of oxygen produced in a biogenic way. The isolation of free oxygen in the stratosphere under the influence of ultraviolet radiation and in connection with decomposition of water vapor and perhaps carbon dioxide, might prove to be much more important. This field of phenomena is still less studied and recorded than the isolation of oxygen in the metamorphic envelope. But two circumstances that greatly diminish the geological significance of this phenomenon should be taken into consideration:

1. The small mass of the rarefied gases in the stratosphere and higher.
2. Their slow exchange with the troposphere.

Finally, there is a third factor to be considered: the dissociation of water molecules under the influence of alpha- and partly beta-radiation of the ubiquitous atoms of radioactive elements. The existence of these phenomena is certain, but nowhere do the concentrations of such atoms seem large enough to be taken into account within the limits of the biosphere. Unfortunately, this phenomenon is not sufficiently studied either through experiment or natural observation.

In view of all this, we can assert now that the free oxygen of the troposphere and the surface waters (gases dissolved in natural waters; i.e., more than one-fifth of the troposphere) is created by life. Furthermore, a quite similar phenomenon is observed for the free nitrogen of the troposphere, and it will be correct to conclude, and to take into account from now on, that the terrestrial gaseous envelope – our air – is created by life. Thus, the history of free oxygen turns out to be a clear measure of the geological and geochemical significance of life.

6 living matter

Life manifests itself in the Earth's crust in a way that is different from the phenomena studied by biologists. Here we notice two new features of its structure. We see that life acts only by means of the energy, quantity, and composition of the matter inherent to it. Secondly, we see that individual organisms move to the background in regard to the greatness of the observed phenomena; we notice only the general, total effects of their activity.

The geochemical manifestations of life present a picture quite opposite to that imagined by biologists and clearly expressed in Cuvier's definition of life more than 100 years ago. The effect of organisms on the migrations of elements of the Earth's crust has almost completely moved to the background, but the matter of the organisms, the motion of its molecules and its energy, manifest themselves in all the observed phenomena. Such a manifestation of life is as real as the richness and complexity of morphological and physiological processes that are studied by biologists as the only reality. Suggesting a new standard of studying life that is completely different from the usual one, we approach unprecedented phenomena and prospects. The complex effects of minute phenomena, which have not attracted the biologists' attention up till now, reveal an unexpected scope.

In geochemistry, life manifests itself through the joint activities of myriads of living organisms. In this totality the statistical laws and generalizations connected with life are studied. Only some separate properties of life attract the mind. In order to be able to study life in geochemistry, it is necessary to present it in the same terms, with the same logical parameters, as other forms of existence of chemical elements to which we are comparing it here; that is, minerals, rocks, magmas, water solutions, and dispersions. In other words, the totality of organisms must be expressed only from the standpoint of their mass, their chemical composition, their energy, their volume, and the character of the space corresponding to them.

Expressing the totality of organisms in these parameters, we should introduce new concepts, new terms for denoting life. I will refer to the totality of living matter expressed in mass, chemical composition, units of energy, and in the character of space related to it. With such an expression of life, all the aggregates of various organisms are fully preserved and have a precise definition: the species, subspecies, genera, etc., that are identified by biologists on the basis of the study of their morphology or physiological functions. The average mass, the chemical composition, the geochemical energy, and the character of space (for instance left or right-handedness) turn out to be as distinguishing for organisms as the characteristics underlying biological classification. In this respect, a geochemist should distinguish between homogeneous living matter comprised of aggregates of one and the same species, genus, race, etc., and aggregates of heterogeneous matter that consist of organisms belonging to different species, genera, races, etc.

The totality of all the living organisms of our planet constitutes living nature. It consists of heterogeneous living matter, which differentiates itself on the surface of continents and islands and in salty and fresh waters, in more or less distinctly separated masses made up both of homogeneous and of heterogeneous living matter. Living matter expressed in mass, chemical composition, energy, and character of space may be studied by geochemistry as well as rock formations and minerals, and compared with them in its manifestation. Here a real analogy exists. Homogeneous living matter corresponds to minerals of plain rock formations, and heterogeneous living matter may be considered as aggregations of a variety of homogeneous living matter (mineralogically diverse rock formations).

The forests of our latitudes, the fields of cereals and other plants, the oyster and other shoals where only one species of living organisms reigns supreme, and the moving herds of animals of the same race, present examples of homogeneous living matter. On the other hand, large tropical forests, herds of animals of different races, great aggregations of life in rivers like the Amazon, the Orinoco, or the Ob, and infinitely diverse biocoenoses [communities] of flora and fauna illustrate heterogeneous living matter and may be compared to mineralogically diverse rock formations.

There is an obvious morphological difference between living matter and rock formations, even though they are expressed in the same parameters. The elements of living matter – the organisms – are always more or less free, often separated from one another and dispersed. The elements of rock formations – the minerals – are almost always inseparably connected with one another, although in sands and in disintegrated mountain rocks the elements of the whole are present as fragments, lie close to one another, and are contiguous. In living matter these elements can be completely distant from one another, dispersed in large spaces, and still comprise one and the same body from the point of view of geochemistry, since their mass, chemical composition, energy, and character of space may be distinguished as such in the surrounding nature.

The organisms expressed as heterogeneous living matter are very poorly studied, and at present they can seldom be distinguished from one another exactly and definitely. But it is clear that the order of inevitable phenomena is one and the same both in living matter and in rock formations. Further on, I shall return to the comparison of their elementary chemical composition, but no special research is needed to understand that, from the standpoint of chemistry, various kinds of living matter differ from one another no less than rock formations. For instance, the difference between the average composition of plankton rich in diatom algae, the composition of living coral constructions, and the composition of submerged thickets of lime algae, is as great as that between marble and quartzite.

The same concerns the masses made up of living matter; they can be compared with the rock formations of the upper layers of the Earth's crust. We often forget these real facts, since we are not

accustomed to seeing organisms as an indispensable and inseparable part of the mechanism of the Earth's crust. Here is something
to illustrate my thoughts. At the end of the last century (in 1889)
the English naturalist, Dr. G. Carruthers, observed a phenomenon
above the Red Sea, which occurred each year on a grand scale; the
migration of locusts from the coasts of North Africa to Arabia. He
estimated the number of these orthoptera in one of the clouds that
rushed over him during the entire day of November the 25th. The
area covered by such a cloud was equal to 5967.3 km^2 (2304 square
miles), and its mass totaled to 4.40×10^7 t. But that was not even the
largest cloud: Carruthers had seen much larger ones. To get an idea
of this number, we can express it in a different way. It is of the same
order as that of the quantity of copper, zinc and lead produced by
mankind throughout a whole century (4.40×10^7 t). At the same time,
the famous entomologists, D. Sharp and N. Ya. Kuznetsov, discussing
the results of Carruthers' observation, saw nothing extraordinary in
them, even with the acceptance of an erroneous value (4.35×10^{10} t)
of the mass of the cloud given by Carruthers, that exceeds the figure
following his observation by many times.

This cloud of locusts, expressed in chemical elements and in metric
tons, can be considered as an analogue to a rock formation, or more
precisely, to a moving rock formation possessing free energy. In
comparison with the variety and extraordinary grandeur of living
nature, a cloud of locusts is a minor and insignificant fact. There
are phenomena that are much greater and more powerful, such
as the constructions of corals and lime algae that are continuous
over thousands of square kilometers, living films of ocean plankton,
floating algae of the Sargasso Sea, the taiga of Western Siberia, or the
hyle (Greek word for forest) of tropical Africa. All these are separate

facts among a multitude of phenomena of the same order. Such masses of living matter can be freely compared to many rock formations, although not to all of them. Large batholiths, or huge layers of crystalline schist, make up conglomerates in the Earth's crust that weigh incomparably more than any aggregate of living matter, more even than living nature as a whole, but these huge masses are outside the realm of living nature. In the biosphere, the aggregations of rocks and living matter constitute masses of a similar order.

In the biosphere, the rock formations, the mass of liquid waters, and the troposphere may be compared to the aggregations of living nature not only in mass, but also in the character of the volume occupied by it. Although the mass and volume of living matter in the world ocean, or the hydrosphere, are very insignificant in comparison to the volume of its inert matter, taking into consideration the incessant movement inherent to living organisms, the importance of living concentrations and their separate dispersions increases enormously. The difference here between living matter and inert matter becomes even more considerable than in the solid crust of weathering. To a lesser extent it concerns the troposphere, of which the only inhabited part is that adjoining the geoid.

The space covered by life (i.e., the volumes of living nature – living bodies) has one distinct peculiarity when compared to the surrounding life of inert matter. This peculiarity was with the anticipation of genius discovered by L. Pasteur in 1838–1848, who named it "dissymmetry of the space of life." It manifests itself in our everyday life in left- and right-handedness, structures, movements within living bodies, and in exposures of living organisms in their environment. I cannot expand on this profound distinction between the manifestation of life and non-life on our planet, which unfortunately has attracted very little attention from naturalists and thinkers up until now.

Recently I happened to speak about the significance of this phenomenon several times. Here it is vital to note that in full accordance with the character of the masses of living nature in the biosphere, the areas of dissymmetry (living nature), and the areas devoid of dissymmetry (the inert matter), are commensurable. The biosphere, the realm of life, is a mosaic – it is comprised of different kinds of space. Life's remains – some organogenic minerals and rock formations – retain the manifestations of dissymmetry for a geologically long time. Dissymmetry without the influence of life is not known on our planet.

From the very beginning of the geochemical study of living nature, two extremely important phenomena attract our attention, which present a distinction from inert matter that is as important as dissymmetry. First is its striking variety. Every species of animal and plant corresponds to some particular homogeneous living matter; in general, the number of described species of organisms exceeds a million. In 1929, R. Hesse counted 1,013,773 described species of animals, and in 1918, M. Constantin corrected the indications of P. van Tiegem by counting 175,300 species for plants. These numbers increase by several thousands every year, and they are still far from being final. Insects alone include millions of different species. It should be accepted that several million kinds of homogeneous living matter exist at present, of which only a small part is familiar to us. These kinds of matter contain millions of specific molecular structures and various chemical compounds that we still know little about. Inert matter is distinctly different from this living matter of the Earth's crust with its striking variety; we know fewer than 2500 species of minerals, and every year this number only increases by a few dozen.

We are facing a phenomenon of great importance here, which is expressed very clearly. It proves the great chemical distinction between living and inert matter in the ways they manifest themselves in the chemical organization of the Earth's crust. In inert matter, only some exceptional phenomena such as lavas and gases may be comparable to heterogeneous living matter in their changeability. This chemical complexity and changeability are a distinct manifestation of the chemical activity of living matter. Any living matter, even homogeneous matter, presents a totality of intense "molecular whirlpools," as G. Cuvier put it. Any organism constantly and irrepressibly captures the radiant energy of the Sun, in a direct or indirect way, and turns it into free chemical energy. A considerable part of the Sun's radiant energy reaching the Earth's surface is thus captured and turned into a new form. Of special geochemical importance is the capture by green plants. The geochemical consequences of this structure of the Earth's surface are truly grand.

These consequences acquire even greater significance thanks to the unique distribution of the matter that is present on the surface of our planet in an active state. Its distribution is, as we shall see, familiar to us, but we fail to realize its importance. Living matter is distributed more or less continuously all over the Earth's surface – it makes up a thin, but continuous cover in which the free chemical energy produced from solar energy is concentrated. This layer is the terrestrial envelope that was named "biosphere" by the famous Austrian geologist E. Suess about 60 years ago, and that constitutes one of the most characteristic features of the way our planet is organized. Only in this film is the specific form of existence of chemical elements concentrated, which we have called 'living matter'.

7   matter in a state of dispersion

All forms of terrestrial existence of chemical elements, such as rock formations and minerals, liquid or semi-liquid magmas and dispersions, as well as living matter, change with different geospheres and terrestrial envelopes. Once more I shall dwell upon the dispersion of chemical elements, because this form of existence is usually ignored. The dispersion of chemical elements has been known for a long time, but its significance has not been sufficiently evaluated by scientists up until now. Its study is an almost virgin area of science, and it is very probable that in the nearest future our ideas about this form of existence of atoms will undergo a crucial change.

There are areas in the Earth's crust in which dispersion of chemical elements is the dominating, most characteristic form of existence. In a vertical section of the planet, two upper geospheres present such areas: the stratosphere, which last year[21] was reached by the Belgian scientist Piccard and his assistants for the first time, and the ionosphere, which is the field of free atoms, ions and electrons. These layers of extremely rarefied atmosphere – "the radiant matter" of W. Crookes (1879) – are sources of energy that seem more and more significant each year. In 1932, these geospheres with their dispersed matter, their manifestations of material radiation of particles smaller than atoms of mainly extraterrestrial origin, showed us

---

21. This was 1932. –*Ed.*

a source of energy that can be mastered by Man, but which we could not have dreamt of several years ago.

Few thinkers of the past, including D. I. Mendeleyev, foresaw the significance of these outer areas of the planet filled with dispersed matter, of the physical vacuum occupying these spaces. These thinkers, and Mendeleyev in particular, expected that the study of the higher layers of the free atmosphere – this "laboratory of weather" – would bring explanations to many riddles in the phenomena of our planet that surround us. Reality suddenly surpassed the most daring scientific expectations.

Everything shows that in this dispersion of atoms throughout the planet's spaces bordering on the cosmic vacuum we see the pure manifestation of the states that we meet everywhere in the biosphere, and that, with attentive scientific consideration, we are just discovering. It permeates the terrestrial matter of the biosphere in the same way way it permeates the "empty" space of the gaseous masses of the stratosphere and higher. How deep do these manifestations reach into the Earth's crust? What is the state of all these elements, the more than 50 elements we find in every drop of salty ocean water, the at least 30 elements of every particle of white Carrara marble, which seems chemically pure to us, the 43 chemical elements found in a piece of a Mansfeld copper schist from Germany, or the 50 chemical elements contained in every drop of Vichy mineral water from France?

What is the state of "traces" (atoms) of Bi, Mn, Cu and Zn, which by means of luminescence phenomena, as proven by P. Lenard and W. Clatt, can be seen in every sulfurous calcium made from calcium of any mineral and all rocks containing lime, even if this lime seemed pure after chemical investigation? And what about the gases of the atmosphere in which, for instance, each cubic centimeter of the air surrounding us contains an atom (a bit more) of heavy radon gas, which irrepressibly turns into still smaller particles (atoms)? What is their state? We know almost nothing about it, but such facts have been accumulating for a long time. Even after putting aside the phenomena of the upper layers of the atmosphere – the mysterious phenomena that arouse our imagination – their significance seems enormous to us.

In the late eighteenth century in Paris, G. Rouelle and J. d'Arsay-ainee proved the dispersion of gold, as was anticipated by chemists such as the Swede, U. Jerne – "the last skald" – in the late seventeenth and early eighteenth century. This idea had originated under the influence of the persistent experimental work of alchemists. H. Davy indicated the general dispersion of sodium in the early nineteenth century. It was noticed but misunderstood by G. Talbot, and in 1859 it was finally seen, together with the dispersion of lithium and calcium, as the result of the basic experiments of G. Kirchhoff and R. Bunsen.

I have already mentioned the fact that, by this time, the dispersion of iodine had been proven but not accepted by Chatain. A series of analogical data can be found in the archives of science, for instance concerning platinum. In many cases, dispersion could be explained by the presence of mechanical admixtures. But the scope of the phenomenon makes it necessary to admit the existence of the unnoticeable transition of mechanical dust, containing for instance sodium, into its atomic dispersion. I. Vorchhammer recorded the same phenomena for the chemical elements of seawater in the middle of the last century.

After several great discoveries of the nineteenth century, such as spectroscopy, luminous phosphorescence, ions of rarefied gases, and especially the phenomena of radioactivity, the dispersion of elements acquired an unexpected significance in our century; it turned out not to be an ordinary phenomenon, but the most characteristic property of the matter of our planet. It cannot be ignored in geochemistry, even if we cannot as yet give an exact explanation of these phenomena; we know they really exist.

In our analyses, this state of the dispersion of atoms is indicated by their presence in "traces": small fractions of mass percentage. For different elements, these data vary from 10^{-10} % (I, Li) to 10^{-18} % (radon in seawater). The significance of these minute concentrations may also be expressed in a form that speaks more to our imagination. We can transfer the quantities of mass percentage – the tiny fractions of the analyses – into the quantities of atoms contained in 1 cm^3. In this case, 1 cm^3 of seawater contains for instance 2.8×10^7 atoms of radium. In other cases, the number of atoms of an element in dispersion can reach 10^{20} -10^{22} or more – a number that cannot be imagined because of its magnitude. The smallest dispersion we measure (including radon and its isotopes) corresponds to a content of millions of atoms in a cubic centimeter of terrestrial matter.

The impression of the dispersion being insignificant disappears in reality. It is replaced by its opposite – the impression of grandeur. This is a clear example of the relative character of the concepts of big and small. In gaseous and liquid media, the enormous number of dispersed atoms is in constant movement at speeds that also exceed our imagination. This movement accounts for the effects they have (i.e., their "work"), which can manifest itself in geochemical phenomena and indicates their active significance.

We can see it now in the terrestrial field, where their realm borders the planet. In gases and liquids atoms rarely collide and their chemical molecules cannot manifest in a form that is familiar to us. Their effect is mainly connected with the electric fields they induce (in gases) and with their influence on partial, capillary, and colloidal forces in natural liquid masses. The dispersed atoms of radon create strong electric fields in the troposphere over land. In solid bodies, both crystalline and amorphous, they take no part in the crystalline lattice (although perhaps for some amorphous bodies?). According to our present-day notions, they remain outside the lattice in the process of crystallization. Being in chaotic motion, they cannot give us an impression of chemical compounds or manifest chemical properties as they neither collide in liquid or gaseous bodies, nor make up static systems characteristic of solid compounds.

As for the matter of the Earth's crust, the problem is obscured by the fact that we do not know what part of the dispersed atoms enters the spatial lattices of solid bodies. The terrestrial solid bodies of the biosphere and the stratisphere,[22] and probably those of deeper spheres, are embraced by an unseen spongy net of capillary spaces, which permeate all solid matter and which are filled with watery solutions. It is very probable that the atoms of dispersion are concentrated in this capillary spongy mass, and that they do not enter the space occupied by the crystalline lattice. We shall come

---

22. Vernadsky uses the term *stratisphere* to mean the part of the Earth's crust made of *strata*; that is, sedimentary rocks. The term is easily confused with *stratosphere* (the atmosphere above about 6 to 17 km). –*Ed.*

across this phenomenon again when I speak about the dispersion of radioactive elements.

In the future, the state and movement of atoms in dispersion is sure to be studied. Apparently, the dispersed atoms move freely, are not united into molecules, change in charge, and are sometimes (or always?) in a state of transformation (sometimes of obvious decomposition). The energy of elements in dispersion is atomic energy.[23]

---

23. Although Vernadsky anticipated the use of atomic energy in atomic bombs and power plants (see page xxxiv), his use of the term "atomic energy" here and on page 314 is in a different context. From his discussions about the atom, it is clear that he was enamored of the (then) new model of the atom with its proton/neutron nucleus and its orbiting electrons, each in a specific orbit as was just then being postulated by the new science of quantum mechanics (by Niels Bohr, Erwin Schrödinger, and others). Part of this model was the constant motion of "dispersed" atomic-sized particles, their *kinetic energy*. Vernadsky is informing us of this in his discussion of "atoms in dispersion" and their "atomic energy." Thus, Vernadsky's use of "atomic energy" in this context really means what we would now call *molecular kinetic energy*. Albert Einstein had convinced the scientific world of the reality of molecular kinetic energy with one of his 1905 papers, and he had also presented his famous equation, $E = mc^2$, well before Vernadsky was writing in 1933, so Vernadsky (and many other scientists) knew that matter could be converted to energy in nuclear reactions, but that is clearly not the context of his use of "atomic energy" on these pages. –*Ed.*

carbon and living matter in the earth's crust

1   carbon in different geospheres and its role

Carbon is one of the most significant elements in the Earth's crust. It is especially significant considering the quantity of its atoms. In the geochemistry of iodine, manganese, magnesium, and silicon, we have seen the role of living matter in the migration of their atoms. Living matter is related to the properties of carbon atoms, and the same influence of life is certain for other elements as well.

The significance of carbon for living nature could be understood properly only after the studies of A. Lavoisier. However, it should be remembered from the very beginning that atoms of *carbon* do not prevail in the organized substance known as living matter, whereas atoms of oxygen and hydrogen do. On average, hydrogen makes up more than half, and three other elements, carbon, nitrogen, and calcium account for more than 1% each.[24] There are organisms that are much richer in carbon (terrestrial plants, land mammals and vertebrates which are approximately 10–20% carbon).

A great many water-rich organisms and all aqueous organisms may contain only tenths of a percent of carbon but usually contain

---

24. Vernadsky here points out that carbon makes up a much smaller mass of organisms than hydrogen and oxygen but his figures can be misleading. Many organisms are about 90 % water, which alone would account for nearly 80 % oxygen, quite a bit "more than half." The implication that most organisms only have little more than 1 % carbon is also misleading. After the water is removed (i.e., dry mass remains), most higher plants consist of 6 % H, 45 % C, 45 % O, 1.5 % N, and 2.5 % total of all other elements. Thus, a typical plant including the water is about 4.5 % carbon. Vernadsky does note that "terrestrial plants, land mammals" may have as much as 10–20 % carbon. –*Ed*.

several percent. From the geochemical standpoint, living matter is oxygen matter rich in carbon, and only at times is it a carbon organism containing more than 10% by weight. The significance of carbon in living matter is not explained by its quantity, but by its chemical properties (i.e., the specific structure of its atoms), and maybe *not only* regarding surface electrons. The dominating, special significance of carbon atoms is characteristic not only of living organisms. The significance of carbon in geochemical reactions is always greater than could be expected with regard to its mass.

The average quantity of carbon in vadose, phreatic,[25] and juvenile areas of the Earth's crust corresponds to a few tenths of a percent. According to the latest calculations of petrographers and chemists such as F. Clarke and Washington (1924), it is less than 0.1 % (0.087 %). I think that these figures do not correspond to reality. They concern only the carbon within carbonates (in the form of carbonic acid) and in igneous rocks, and even in the latter the figures concern nothing but the carbon dioxide of an ordinary chemical analysis. The igneous rocks, in which only this kind of carbon is considered, make up about 95 % of the whole Earth's crust. The analyses however, ignore both the gases rich in carbon and the organic matter poor in oxygen, which are always contained in these rocks.

A large quantity of carbon compounds is concentrated in metamorphic rocks such as gneisses. Those rich in carbon (for example graphite gneisses) are much more widespread than is usually supposed. The compounds can be found in paragneisses and in orthogneisses. Granites and acid-massive rocks containing graphite often form large

25. As *vadose* refers to Earth above the water table, *phreatic* refers to the ground water below the water table. Phreatic can also refer to explosive volcanic activity involving steam derived from ground water. –*Ed.*

massifs. Unfortunately, ordinary chemical analyses ignore this carbon, and our information is based on data that should be checked with the new methods of chemistry. In determining the average composition of igneous rocks, large quantities of paragneisses and corresponding granite-gneisses are ignored. These are always richer in carbon than granites and igneous gneisses. A deeper study of volatile substances, which can emanate from all igneous rocks and gneisses when they are being heated, would be of great interest. All this carbon is ignored by Clarke and Washington. In their calculation of the average content of carbon, they also failed to take into account the carbon of the sedimentary and upper metamorphic rocks. Carbonaceous shales for instance, which are related to coal deposits, contain a larger mass of carbon than coal do. The Proterozoic (Algonquian) carbonaceous shales contain sometimes more than 20% carbon.

Juvenile and phreatic, gaseous and solid carbides such as carbonols, carbon oxide and maybe cyanides, are completely excluded from these definitions, as well as the graphite of the deeper layers of the Earth's crust. Gaseous compounds of carbon permeate all substances of the Earth's crust. They are not taken into account either in inert matter or within living matter. Even if these gaseous parts are of no significance due to the little influence of even the total carbon in living matter upon the calculation of carbon in the Earth's crust, the gaseous compounds of the Earth's crust as a whole cannot be ignored. Thus, the quantity of carbon in the Earth's crust is surely larger than that given by Clarke and Washington.

According to my earlier calculation, the average content of carbon in the Earth's crust must be about 0.4 – 0.5 % of its mass. Carbon seems to be less widespread than titanium (more than 0.6 %). It is extremely uneven throughout the Earth's crust. We see a very

definite concentration of carbon in the sixth and seventh chemical geospheres, which unfortunately we cannot present in figures
as yet. For sedimentary rocks, the content of carbon in the form of
CO_2 and in carbonaceous matters approaches approximately 2%. In
fact, divergence of carbon content from this average figure are quite
pronounced. In countries rich in limestone, the quantity of carbon
in the parts of the crust with a prevalence of rocks containing this
element (such as coal, bitumen, and carbonates) may rise to 10–
12 %. In the biosphere it can be larger still if oceans are excluded. In
the parts of the crust consisting mainly of sandstones, clays, igneous
rocks, and crystalline shales, and in the upper geospheres in the
ocean, its quantity sometimes falls far below the average figure.
But in the upper geospheres of the lithosphere, a certain increase
in carbon content is generally observed, sometimes exceeding the
average quantity of the Earth's crust by five to six times.

With the increase of carbon, geochemical processes increase much
faster than in proportion to its mass. There are two important
phenomena to be especially noted in geochemical history:

1. Formation on the Earth's surface of living matter – the most
    characteristic element of which is carbon – and the dominating
    gaseous form of deep carbon minerals. These two facts make an
    indelible imprint upon the whole chemistry of the Earth's crust.
2. The stability of carbon minerals.

On the other hand, it is necessary to take into account the specific
properties of carbon. The most characteristic property is the stability of its organic natural compounds in terrestrial thermodynamic
conditions. They change very slowly there, and even nitrous carbon
compounds, which we used to consider very unstable, often don't

change at all. In crude oil for example, we always observe nitrous compounds with a very stable chinoline structure, and the chitins of certain brachiopoda have been preserved in sedimentary rocks since the beginning of the Cambrian era – for many hundreds of millions of years.

Protein bodies, as well as chlorophyll, are preserved in favorable conditions for many hundreds of thousands, and perhaps even millions of years. In the thermodynamic and chemical conditions of the Earth's crust, all the natural carbon compounds are stable if they are protected from the influence of life. The compounds comprising living matter – proteins, fats, and carbohydrates – remain untouched and change extremely slowly without losing their structure, on the condition of being isolated from the phenomena of life. Rotting and fermentation in the biosphere, changes of living matter after death, are biochemical phenomena.

The well-known experiments of L. Pasteur have indisputably proven the stability of these organic bodies in a sterilized medium, while in our ordinary environment they show a characteristic example of instability. Everywhere in nature we see numerous examples of preservation, or only a slow change of these products of living matter, if they remain untouched by life. This can occur in a lifeless environment, even at the temperature and pressure of the biosphere. In peat bogs and in arid countries for example, the substance of organisms remains unchanged. The most complex molecules of the compounds of which organisms consist change little there. Carbonaceous minerals are still more stable than the immediate products of living organisms, the substances composing their bodies.

In living matter and in organic chemistry, compounds of carbon show definite and numerous changes. Organisms contain millions of different compounds (a large organism probably contains tens of thousands), and synthesis in our laboratories has created hundreds of thousands. In these laboratories, they show various molecular transformations and form new compounds with distinctly different chemical properties. Carbides existing outside living matter behave quite differently. They also seem very numerous, but we can hardly distinguish them from living matter. In ordinary reactions of organic chemistry, we have the same humic or stable resinaceous products as products of secondary reactions, but we do not study them, and we ignore them while calculating the output of the reaction. We cannot do this in biogeochemical or geochemical processes. Here we must study *all* newly formed bodies, even if they are alien to the main reaction.

2   the stability of carbon minerals

Everywhere in living matter we can observe the formation of such stable and inert carbonaceous bodies. They take part in the formation of the most stable and strong parts of organisms, and are still more frequent and numerous in the waste and excrement of living matter. They concentrate in its remains after death and persist in carbonaceous minerals. This is understandable, since almost all carbonaceous minerals have formed biochemically or take origin in the remains and wastes of organisms. Their quantity increases over the course of time, as organisms constantly form inert bodies that decay very slowly. The scientific study of these products is beginning to make slow but steady progress. It is evident that the development we are now seeing in these new chapters of organic chemistry, which is revealing the structure of extremely stable and inert carbon molecules, will be as rich as that of previous chemistry. The study of these compounds is important for the chemical history of carbon, and we will understand the properties of this extraordinary element only when we take into consideration all its molecules: stable and labile as well as inert and active.

For carbon, three different forms of existence should be noted. All of them play an important part in its history. These are:

1. Millions of chemically different compounds of carbon that are stable only in the biosphere (within the limits of an organism during its life).
2. Thousands of products of waste and exhalation of living matter, which little by little turn into carbon minerals and thus affect the chemical balances of life.
3. Hundreds of millions of carbon minerals that have appeared as a result of terrestrial chemical reactions outside living matter.

The slow and incessant transformation of organic wastes and living matter into vadose minerals of carbon upon death is the central phenomenon in the geochemical history of carbon and of such important vital elements as O, H, N, S, and P.

3 dispersion of carbides

This process results in the dispersion of carbides, permeating all the matter of the biosphere with carbon. The remains of organisms are dispersed everywhere in minute traces – in waters, soils, precipitation, sea silt, and all minerals appearing in the biosphere as the result of any process. By means of geological processes, the substance of the biosphere leaves it and penetrates into deeper envelopes of the Earth. This explains the ubiquity of biogenic carbon or carbonaceous remains of living matter in these regions. *The chemical nature* of these carbon traces is not always exactly known, but everything leads to the conclusion that except for carbonates, they comprise different carbonaceous bodies that little by little turn into a stable state of free, non-fixed, elementary carbon. The analysis of this geologically perennial process shows that in respect of eons of geological time, the initial body for all intermediate organogenic compounds will be elementary free carbon. Here we have a closed circle.

Probably two different groups of such stable, organogenic, dispersed intermediate compounds exist. One group contains oxygen compounds such as *humus substances,* and the compounds of the other approach *hydrocarbons.* Both turn into native carbon, but hydrocarbons in vadose and phreatic layers are very stable and are preserved for a geologically long time. The chemical character of these traces is in genetic relation to the structure of the carbonaceous compounds from which they originate. This dispersion of carbon only partly resembles dispersion of other elements such as iodine or bromine, because the final product of this dispersion (elementary carbon, graphite) is a solid body, which does not exist in a liquid or gaseous state under the conditions of the Earth's crust; nothing shows emanation of carbon *atoms* here.

This dispersion of carbon has an outstanding significance in its geochemical history. Two facts explain its mechanism: the properties of living matter and the existence of numerous gaseous compounds of carbon that are often very unstable. Traces of carbon that are found everywhere have different origins. On the one hand they are the last remains of organisms, microscopic and ultrami-croscopic, and on the other hand the final products both solid and liquid – of the decay of various carbonaceous molecules that now exist in a state of extreme dispersion, such as in terrestrial gases, for example.

On changing upon death, organisms release gaseous, liquid, or solid carbonaceous products. This phenomenon is unchangeable for all organisms: large and small, visible and invisible. But in the remains of green terrestrial plants and large algae, a significant role is played by *humus compounds*, while hydrocarbon products prevail in the remains of animals, plants without chlorophyll, unicellular algae, spores, and seeds. The smallest organisms are of the greatest impor-tance for the dispersion of carbon.

The existence of minute microbes and ultramicrobes arouses profound questions as to the matter that forms them, and these problems are often forgotten. Each of them contains tens, if not hundreds, of various chemical elements [compounds], and at the same time each is distinctly different from the surrounding matter. They are centers of chemical activity in the form of limited masses of liquid or solid matter in a state of extreme dispersion. The small-est organisms approach large molecules in size.

Living matter is one of the most powerful forces of dispersing matter in the Earth's crust. The Earth's crust has no other force equal to it. Dispersed matter is always the most chemically active – a fact well known to us through our laboratory practice. In order to disperse some substance, an effort must be made to break the cohesion of its particles. Even for the minute particles of liquid and solid matter, there are limits that cannot be crossed without their decomposition, limits connected with the forces of chemical affinity, cohesion, and surface and capillary forces.

Friable, powder-like dispersed matter can exist only on the surface of our planet. In the course of time it slowly but steadily becomes compressed and coalesces. It is not present in the upper metamorphic envelope, since the effects of pressure and chemical and colloidal processes compress all the dust into solid masses. On the Earth's surface, the forces that disperse matter are decomposition of gaseous products and the simultaneous formation of solid or liquid coatings, volcanic eruptions, solid or colloidal products of liquid spray, erosion due to the effects of water and wind, and finally, *living matter*. Living matter is the most powerful of all dispersing agents, the only one that is constantly acting *everywhere* and ultimately disperses the greatest mass of living matter.

Living matter forms special "dust" – microbes insignificant in size, but possessing a very intricate chemical composition. They manifest great chemical activity and have an absolutely unique ability to quickly increase the quantity of similar "indivisible specks of dust." The capacity of organisms to multiply is a great force changing the structure of the biosphere. This manifests most distinctly in the influence of organisms upon the migration of atoms, in other words, in their geochemical energy. For microbes, such a manifestation is the rapid growth of the mass of living matter (the smallest "specks of dust"), increasing in geometrical progression in the course of time, which leads to an equally rapid "dispersion of matter in the biosphere."[26]

---

26. On geochemical energy and proliferation of organisms, see Vernadsky, V., *The Biosphere.*

The size of living organisms sometimes reaches the size of dispersed particles (micelles of colloids). No doubt, they approach the order of 10^{-5} cm, or even 10^{-6} cm, but cannot reach the size of atoms and molecules (10^{-7}–10^{-8} cm.) These smallest living bodies exist in the microscopic aspect of the cosmos, in the realm of particle forces. "Dispersing" matter by its multiplication, life reaches the physical limit of mechanical dispersion and approaches the phenomena of creating molecules, i.e., the deeper dispersion of matter in the given medium by means of chemical forces.

Microbes and other unicellular organisms die and decompose into smaller particles only in cases of catastrophes – extreme cases. Usually they do not leave the circle of living matter; they continually decompose within it and do not leave the bodies of organisms. Even at the death of whole biocoenoses [communities], they usually do not give friable dust but stay together in solid "earthy" masses that appear in this process and that are soon taken up by life. The smallest organogenic "carbonaceous dust" appears during the life, and especially after the death of other large organisms, in case these organisms do not serve as material for the life of others.

All these remains, both large and small, are rich in carbon. Some of them make up gaseous products such as CO_2, etc. Another part turns into the smallest solid and liquid carbonaceous "dust." That dust, due to powerful capillary forces proportional to its surface, is absorbed [or adsorbed] and retained by all liquid or solid external bodies such as vadose minerals, water, gases and living matter. The smallest dust floats in gases, makes up colloids (and lees[27]) in water,

---

27. Lees = sediment settling during fermentation, especially in wine dregs. – *Ed.*

and penetrates solid bodies and sticks to them. Eventually it turns out to be present throughout the biosphere's substance, and its total mass is tremendous.

Another source of dispersion of carbon, of formation of some of the smallest carbonic particles, is decomposition of natural carbonaceous gases. Carbonaceous gases generally have a biogenic origin. Living matter constantly emanates an enormous quantity of volatile gaseous products. Partly they are products of respiration and metabolism: "smells." Especially the infinitely diverse bodies of the latter make up the light, solid and liquid "dust" and "lees." They easily change chemically, making up solid or liquid inert stable products, such as the products of terpenes' oxidation. Gaseous products like CH_4 and others often appear in traces during the decay and decomposition of living matter, in a medium that is poor in oxygen (for instance under water) and rich in the products of life. They can be of juvenile or phreatic origin, and usually rise from the depths of the Earth's crust by way of slow, gradual vaporization.

These gases, dispersed in the surrounding atmosphere, decompose and leave thin lees. For instance, methane and other hydrocarbons decompose under the influence of electric discharges during thunderstorms or alpha-radiation. Solid, dispersed sediments of these gaseous emanations make up films such as the finest coal dust that accumulates on all surrounding bodies. Little by little these remains turn into pure carbon, or in case of the presence of active oxygen, into carbonic acid[28] [i.e., CO_2].

---

28. Carbonic acid is H_2CO_3, which is CO_2 reacted with H_2O; Vernadsky seems to use CO_2 and carbonic acid interchangeably. –*Ed.*

There is almost no doubt that the similar decomposition of gaseous carbides in the form of solid dust takes place also in deeper geospheres, such as in the stratisphere and in the metamorphic geosphere. There are indisputable indications that a number of deposits of graphite (which sometimes accumulates in very small crystals) releases carbonaceous gaseous bodies by way of decomposition that are not exactly known at present, and that are sometimes identified as carbon oxide or metallic carbonyls. But the dust-like products in these envelopes usually crystallize again, and grow and accumulate into large masses. This dispersion of carbon cannot be ignored. We know that over geological time, small phenomena eventually have tremendous effects.

Such dispersed matter gives bodies of diverse and sometimes very intricate structures, which can enter into chemical reactions more rapidly and energetically than more solid or liquid masses of the same mass. On the other hand, the insignificant liquid or semi-liquid sprays or films can accumulate into great liquid masses under favorable conditions. We see the significance of this fact for instance, in the geological history of crude oil, which are natural hydrocarbons. But even when remaining in a state of dispersion, each of these carbonaceous particles turns into CO_2 or graphite after passing through a certain cycle of changes.

The carbon of such dispersions, as well as the carbon of living matter and vadose minerals, eventually originates from compounds of juvenile carbon (the carbon of the deep layers of the Earth's crust), and to a certain extent from native carbon, hydrocarbons, and carbon dioxide. Here we see the same closed circle; the initial and final products are identical. A current of carbonaceous gases, such as CO_2 and hydrocarbons, incessantly flows to the biosphere

from the deep geospheres of the Earth's crust. The gases have originated, as we shall see below, from native carbon.

4   primary carbides

The state of our knowledge about the primary juvenile carbonaceous
minerals is rather deplorable. We lack exact facts, and it is impossible
to present the geochemistry of juvenile carbon in a purely empirical way without turning to hypotheses that are only more or less
grounded. Part of such carbon is presumably related to little known
and little understood processes that take place in the deepest parts
of the Earth's crust, for instance under the granite geosphere or at its
lowest border; this has something to do with magmas.

Such a primary origin independent from the biosphere was recently
ascribed to the largest accumulations of petroleum and bitumen by
the most outstanding researchers. I think that the facts contradict
such an idea, but it still has followers. At the end of the nineteenth
century, the non-organic juvenile origin of oil deposits was admitted
and enthusiastically supported by scientists such as M. Berthelot,
H. Abich, D. I. Mendeleyev, and H. Moissan – they all supported
these views until their deaths (i.e., up until the first decade of the
twentieth century).

Under conditions of high pressure and in the presence of magmatic
masses, formation of many oil hydrocarbons is possible, but everything we know about the conditions of the bedding, the chemical
composition, the physical properties of petroleum (in which hydrocarbons only dominate, but do not totally comprise them), and
their large deposits, decisively contradicts this idea of their genesis,

partly because petroleum does not consist of hydrocarbons alone. Of course, we cannot exclude the possibility of a juvenile origin of special forms of hydrocarbons different from oil, for instance in some crystalline rocks, but no confirmed cases exist. I shall expand upon this issue below. Here I shall present the exactly stated facts concerning the primary (i.e., independent from the biosphere) minerals of carbon.

It should be borne in mind that such primary juvenile minerals of carbon may form in various upper geospheres up to the biosphere, as for instance in the case of magma penetration. As was mentioned before, until now we have not been able to study the rocks of the magmasphere – the granite biosphere – precisely. We study the juvenile, primary carbon minerals in the biosphere in the form of magmatic minerals connected with igneous rocks, and transfer our ideas received in this way to the granite geosphere. These primary compounds correspond to the carbon close to the surface of the geoid; all of them form within the limits of the Earth's crust.

Such primary emanations of carbon are divided into two distinctly different groups: *oxidized bodies and bodies containing no oxygen.* Oxidized bodies, in their turn, can be divided according to their content of oxygen. The first group contains carbonates, carbosilicates, and coal silicates with a carbonate group in the side chain. All of them are in a genetic relationship with carbonic acid. Carbonic acid makes up either salts for primary bodies – mainly potassium ones – or forms products that join with aluminosilicates, for example: $3Na_2Al_2Si_2O_8 \times Ca (HCO_3)_2$. There are igneous rocks containing up to 1.7% CO_2 (= 0.74% C). Igneous rocks with juvenile calcite are still richer in carbon: Trachyte from Bilbao contains 7.69% CO_2 (= 2.09% C), and phenote from Norway 35.2% CO_2 (= 9.6% C).

The second group, poor in oxygen, includes CO, CSO, HCHO (formic aldehyde), and HCOOH (formic acid). Usually, these are molecules that form at high temperatures, following the reduction of carbonic acid in the presence of water and hydrogen sulfide. They are not very rare, but they appear only as traces in juvenile and phreatic gases. No doubt, thiocarbonic acid and derivatives of formic acid play a very important role in the chemical phenomena of the Earth's crust. It is possible that carbon oxide is to a great extent independent from carbon dioxide in regard to its genesis. There are indications that carbonyl compounds of iron and nickel may be present in the Earth's crust. Among the carbonaceous minerals containing no oxygen, of greatest significance are the hydrocarbons CH_4, C_2H_6 etc., metal carbides, and native carbon.

Evidently, the chemical conditions necessary for the formation of the two groups of carbonaceous minerals – oxidized and containing no oxygen – are incompatible with each other. Their existence may indicate that they originate at different depths of the magmosphere. Experiments have shown that hydrocarbons indeed form under such conditions. At the same time, by analogy with meteorites, the presence of metal carbons in the Earth's depths is often suggested. The existence of petroleum was considered a manifestation of their decomposition. A deeper study of these phenomena has completely disturbed this too simple assumption, and it has illuminated problems in the field of juvenile carbon and its primary compounds.

It turned out that all juvenile minerals of carbon that are rich in oxygen can in any case be related to CO_2, since carbonates decompose at the temperature of the magmatic geosphere, and possibly even at that of the deeper areas of the metamorphic geosphere.

Aluminosilicates containing carbon are *secondary* juvenile and phreatic products. They form by the effect of carbon dioxide on existing aluminosilicates in kaolin, and in all probability mainly by its effect on feldspars at high temperatures. Carbonate-silicates are very rare and include only an insignificant number of carbon atoms.

In deep phreatic and juvenile geospheres, carbonic anhydride is very stable and probably very abundant. But in these parts of the Earth's crust there are not appropriate conditions for its synthesis, only for dissociation of previously formed carbonates. We know that in the deep layers there is no free oxygen. The small quantities that are found in the gases of mineral springs and gas streams coming from that region are probably of different origin. At the same time, there are numerous vadose and phreatic chemical processes, often related to decomposition of carbonates, that produce carbonic acid. Very dispersed carbonaceous bodies that are rich in carbon, or consist of pure carbon, oxidize in the presence of free oxygen and produce CO_2. This process takes place only in the upper phreatic and vadose areas. These compounds, having reached deeper layers containing no free oxygen, dissolve in liquid or viscous magmas and crystallize again as graphite or diamond, or perhaps vaporize as hydrocarbons.

Carbonic acid, which is isolated in great quantities during volcanic eruptions and in volcanic areas, is created not only by means of juvenile synthesis, but at least partly by decomposition of existing carbonates resulting from high temperature and to the melting of metamorphic and vadose rocks. These rocks, including limestone, melt gradually as they descend to the deeper layers. The same phenomena are observed when rocks containing carbon and carbonates are captured by magmas ascending to upper geospheres. The carbonic acid of volcanoes and thermal springs is undoubtedly juvenile (i.e., originates from juvenile areas). But its eventual origin is in essence vadose or phreatic, as was proven by R. Delkeskamp for thermal springs. Carbonic acid is stable in the deepest areas of the Earth's crust but is not always a product of magmatic reactions. This phenomenon is confirmed by the fact that the masses of volcanic carbonic acid are comparable in size with masses of carbonic acid that could be produced by means of decomposition of vadose or phreatic carbonates.

The quantity of carbonic acid of "juvenile" origin that reaches the Earth's surface is enormous. I. Boussingault was the first to note the importance of this phenomenon for the balance of carbonic acid in the atmosphere. Carbon dioxide often dominates in volume but not in mass within volcanic gases. Sometimes it can be observed in an almost pure state. For instance, F. Fouquet found 95.37% CO_2 in the gases of the Santorin volcano, and C. Sainte Claire Deville found 97.1% CO_2 on the isle of Panarea (the Lipar Islands). Carbon dioxide dominates in volcanic gases of the ancient volcanic areas, where its emission is the last stage of earlier eruptions. In the gaseous emanations of active volcanoes, it usually plays a secondary role. In the ancient volcanic areas of the Tertiary, for instance in Europe – in the Eiffel, Rhön [Mountains], Vogelsberg, Sicily, Italy, as well as in the Caucasus – gaseous fumes of carbonic acid have formed due to continual outgassing over millions of years.

The same fact is observed on other continents, for instance, in Asia, North America, and New Zealand. Boussingault calculated that the quantity of CO_2 released by Kotopachi during one year exceeds 10^9 m³; that is, it exceeds the quantity of carbonic acid annually released by the life processes and factories of Paris (70 to 80 years ago this was 3×10^6 m³). Carbonic fumes in the region of ancient volcanoes of the Tertiary give annual quantities that are incomparably larger. The total quantity of such carbon dioxide in the biosphere is probably of the same order as that of carbonic acid in the hydrosphere and atmosphere.

Not smaller is the quantity produced by thermal springs, tens of thousands of which are concentrated in areas of the biosphere reflecting movements of the Earth's crust. Gas flows of nitrogen and methane, which are not connected with new or geologically recent volcanic phenomena, always contain CO_2, but apparently it is never a leading component. Unfortunately this issue needs checking, but apparently there are no carbonic streams or fumes with a prevalence of carbonic acid in these areas.

All coal or oil deposits are observed in modern or geologically recent volcanic areas. The existence of juvenile carbonic acid in these areas shows itself in other phenomena as well. A lot of vein deposits related to geologically ancient and geologically new tectonic movements indicate the participation of carbonic acid in their formation. Part of that carbonic acid has been included in the smallest microscopic pores of the vein minerals and in the minerals themselves.

The same phenomenon takes place now in ancient and new rocks, in particular in granite rocks. Huge quantities of carbon dioxide are concentrated in a dispersed form. Especially the quartz of granite rocks contains carbon dioxide as microscopic inclusions in a liquid or a gaseous state. The quantity in which it is present is enough to explain, for instance, the emission of carbonic acid by thermal springs in some areas of West Germany, as was shown by Laspeyres. He calculated that 1 km³ of granite in the Rhine provinces contains 9×10^{11} dm³ of gaseous CO_2. It follows from his calculations that the total quantity of carbonic acid (liquid) in granites is greater than that in the atmosphere.

The emission of carbonic acid in volcanic areas, some of the thermal springs, and during decomposition of granites, is obviously related to magmatic carbonic acid. Part of this carbon dioxide is undoubtedly of phreatic or vadose origin, and appears when magma melts carbonates, and especially limestone ($CaCO_3$) and dolomite ($CaMgC_2O_6$). Is all the carbonic acid of such origin? Apparently this is not the case. The study of volcanic gases proves that juvenile formation of CO_2 is very probable, but obviously this synthesis is very limited in size and can only explain a relatively small part of the magmatic carbonic acid.

The research of A. Gautier, R. T. Chamberlin, K. Huettner, A. Brun, A. Daya, E. Shepherd, T. Jaggar, and E. Allen on volcanic gases and mountain rocks, as well as laboratory experiments, prove that carbon dioxide can form under the conditions present at a time when magmatic gases come into being. These gases in general can be compared with gases that are released upon heating and melting of erupted rocks in a vacuum, or in an atmosphere devoid of oxygen and water vapors. The gases that are educed in this case, being partly products of new chemical reactions during experiments, were not formerly contained in the rocks; part of them forms during the experiment, following the interaction of water and iron compounds present in the rocks. But, as the melting of the rocks accumulated by the lava takes place during any volcanic eruption, the same gases must be separated in volcanoes too.

The following gases, which occur in volcanic eruptions in great quantities, should be mentioned: H_2O, H_2, CH_4, CO_2, CO, formic acid, and CSO. In addition, it should be remembered that water that is isolated during the melting and heating of rocks, and part of the magma's water, appears due to decomposition of compounds such as aluminosilicates and silicates of the same reabsorbed rocks. Carbon dioxide, which is isolated in magmatic and volcanic gases, can appear and decompose in a series of different balanced reactions, some of which have been investigated experimentally. Here are some of the ones that are most investigated:

$CO_3 + 3\,H_2 \longleftrightarrow CO + H_2O + 2\,H_2$ (at the temperature of red heat)

$3\,CO + 2\,H_2O \longleftrightarrow 2\,CO_2 + 2\,H_2 + CO$ (between 1200 °C and 1500 °C)

$4\,CO + 2\,H_2 \longleftrightarrow 2\,H_2O + CO_2 + 3\,C$ (between 900 °C and 1000 °C)

$4\,CO + 8\,H_2 \longleftrightarrow 2\,H_2O + CO_2 + 3\,CH_4$ (between 1200 °C and 1220 °C)

Also:

$C + 2\,H_2 \longleftrightarrow CH_4$ (between 500 °C and 1000 °C)

$4\,CO + 2\,H_2O \longrightarrow 3\,CO_2 + CH_4$ (between 250 °C and 275 °C)

$CO + H_2O \longrightarrow CO_2 + H_2$

$CH_4 + H_2O \longrightarrow CO + 3\,H_2$

We see that CO_2 is constantly decomposing and reforming during these reactions, and that the molecules that have initiated its formation can also be recreated upon its decomposition. It is possible that CO, CH_4, and other hydrocarbons present in volcanic gases are products of the decomposition of carbonic acid; its oxygen may be used by ferrous compounds. A. Gautier thought that CO originates as a result of CO_2 reduction by the salts of ferrous oxide (a reaction that indeed takes place under these conditions). K. Huettner explained the origin of CO by the action of H_2 on CO_2, a reaction that is certain under these conditions. But CO is possibly a product of the following reaction studied in detail by A. Gautier:

$$4\,CO + 2\,H_2 \leftrightarrow 2\,H_2O + 3\,C + CO_2$$

This is a reversible reaction in which free oxygen takes part that is present in sufficient quantities in magmas and erupted rocks. With the action of water and carbonic acid in the depths of the Earth's crust, great masses of CO can form that inevitably turn into CO_2 on the Earth's surface in the presence of free oxygen as T. Jaggar has shown. CO is present in the atmosphere in insignificant quantities, and it does not accumulate. But the study of these complicated balances gave no indication of juvenile carbon dioxide, if CO capable of giving CO_2 under the influence of water vapors at temperatures of 1200 °C and 1250 °C is not such an indication. *At present, pure carbon (graphite) giving CO and CO_2 as secondary juvenile products can be acknowledged as a primary juvenile mineral.* Passing through CO, graphite turns into CO_2, but the quantity of CO_2 formed in such a way cannot be sufficient to explain the observed phenomena. Possibly, it is one of the reactions of oxygen absorption in the deepest layers of the Earth's crust.

Up until now we remained in the realm of trustworthy facts. We see that CO, CO_2, and CH_4 can form in the Earth's depths at the expense of other juvenile bodies whose presence arouses no doubt (C, CO_2, H_2O, and ferrous compounds). But there are *less certain* facts pointing out that the conditions of the reaction are in reality more complicated. More than once it has been hypothesized that the depths of the Earth's surface may contain quantities of primary carbonaceous bodies, juvenile metallic carbides, and CO capable of engendering large amounts of hydrocarbons under certain circumstances. It was and is still thought that under natural conditions chemical reactions take place that are similar to those studied in our laboratories:

$$CO + 3\ H_2 = CH_4 + H_2O$$

$$Fe_2C + 2\ H_2O = 2\ FeO + CH_4$$

or possibly theoretical reactions:

$$3\ Fe_2C + Fe_2O_3 = 3\ CO + 8\ Fe$$

$$2\ Fe_2C + CO_2 = 2\ CO + 4\ Fe + C$$

Sometimes it is assumed that primary metallic carbides and CO are situated *under* the silicate envelope – in the Earth's metallic core. Indications of their existence there were seen in the large accumulations of hydrocarbons (oil) that were observed in obvious connection with gas-forming movements, and with the tectonic structure of the Earth's surface. But these movements, and the related structure of the Earth's crust are surface phenomena of the planet; water is also concentrated in the surface envelopes. Its presence in the metallic core – if it exists – is hypothetical, and cannot be confirmed by geological phenomena taking place at a distance of thousands of kilometers from this core, in a completely different physical environment.

Still, there are facts proving that metallic carbides, cohenites, and maybe some other compounds are really present in some erupted rocks under conditions that do not prevent formation of hydrocarbons, such as contact with hot water. Unfortunately, this presence has been studied more intensely only in basalts of the Disco Isle and other islands of Western Greenland. Ferrous carbides containing nickel have an obvious connection with metallic iron, and their content is much higher than in meteorites. The carbon of these basalts was explained by special phenomena, such as the capture of lower masses of coal by melted masses of basalt during their ascent to the Earth's surface.

The same carbides can be found in other deposits of terrestrial native iron, and it is possible that a more thorough study of these minerals will show their existence throughout the deep basalts (basalt covers). But even in this case, the formation of hydrocarbons from metallic carbides cannot be of great importance in the history of large deposits of petroleum. The geology and chemistry of oil have made great progress in recent years; the phenomenon seems clear as a whole now. We must consider oil as phreatic minerals forming outside the upper parts of the biosphere, but genetically connected with living matter. Organisms are beyond doubt the initial substance of oil, and participation of anaerobic bacteria in their synthesis is also possible. Oil cannot contain significant quantities of juvenile hydrocarbons.

Before passing on to the genesis of oil I want to say something about *juvenile* carbon oxide. Indeed, this oxide may exist without any genetic connection with CO_2 because of the following reaction studied by A. Brady: $CO + 3 H_2 \leftrightarrow CH_4 + H_2O$, which is reversible like all such reactions, and which shows the possibility of CO formation through the action of water on methane. Carbon oxide is in fact very stable at high temperatures and must accumulate in deep parts of the crust without decomposing.

5 petroleum and its formation

Chemists and geologists studying petroleum usually admit the non-juvenile origin of large masses of crude oil and the solid products it contains, as well as carbonaceous gases that are obviously related to them. It is necessary to stress here that crude oil cannot be regarded only as hydrocarbons. Hydrocarbons *dominate* in the composition of petroleum, but apart from that they always have a large percentage of compounds containing O, N, and S. The explanation of the genesis of crude oil cannot be based only on the explanation of the origin of hydrocarbons, which is often forgotten.

These notions cannot be ignored, because there are still followers of the hypothesis that petroleum is of inorganic origin. Hydrocarbons in this case must form in the deeper parts of the planet, situated deep below the limits of the Earth's crust. This possibility cannot be denied, but considering the present-day state of our knowledge, only suppositions can be made about the physical and chemical conditions in those unknown areas of our planet. It turns out that on the Earth's surface, in the areas accessible to precise scientific study, there are no signs of such hypothetical reactions except the crude oil itself.

The absence of any further traces of such processes would be very improbable if they were really the origin of large oil masses. A deeper study of the chemical composition of natural oil arouses new problems leading to new hypotheses, still less testable and still less probable. That is why it is easier to put aside these complicated constructions analogous to the old epicycles of astronomers, and try to understand the history of crude oil by remaining within the upper envelopes of the Earth's crust accessible to our direct study.

In this case we must consider petroleum as minerals of biochemical origin subjected to a strong metamorphism.

In a new and deeper understanding, we return to the notion about the origin of crude oil, bitumen, and organogenic carbonaceous minerals that was created at the end of the eighteenth century by Buffon. Although the mistakes of this great naturalist and the incompleteness of his ideas about life are quite obvious now, they contain some traits of a deep embrace of nature that have not lost their importance. Two kinds of deposits can be noted for large crude oil accumulations, those found in sedimentary rocks, and permeation of bituminous slates with hydrocarbons. Both types can be regarded as parts of one and the same phenomenon. The accumulations found in slates contain the greatest quantities of crude oil. The origin of oil cannot be explained while ignoring bituminous slates. The hydrocarbons they contain are connected with processes of colloidal character, which I cannot dwell upon here.

Oils are mixtures of different organic compounds in which hydrocarbons dominate. Their composition had been considered to be not very diverse, for hundreds of analyses showed almost the same correlations between C and H for oils of different localities, and of very different historical ages. The content of C in hydrocarbons oscillates between the limits of 83–86 % (sometimes to 87 %). The content of H in hydrocarbons oscillates between the limits of 11–13 % (sometimes to 14%). But finally it became clear that the chemical structure of these hydrocarbons *can* be very diverse.

The first well-studied crude oil, that of Pennsylvania, consisted almost completely of paraffins (C_nH_{2n+2}); from methane (CH_4) to the last liquid paraffin $C_{18}H_{38}$ and beyond. The works of F. Beilstein and A. Kurbatov, and later of V. V. Markovnikov and his students, showed that the oil of the Apsheron peninsula contain hydrocarbons of unknown composition, different from the ones that Markovnikov called naphthenes. These are now known to be very stable compounds with a ring structure and the common formula C_nH_{2n}. In some oil of California, aromatic hydrocarbons such as benzol [benzene] (C_6H_6) and others dominate.[29] Furthermore, there is oil similar to bitumen that is rich in compounds C_nH_{2n} and C_nH_{2n-r}. Now we know that oil consists of a mixture of hydrocarbons, which are very different depending on the geological conditions of their genesis and the composition of the primary organic bodies from which they originate, and which are created by certain organisms.

But in every kind of crude oil there are other molecules, besides hydrocarbons, that had long been ignored. Only in the twentieth century, the American chemist Mabery showed that petroleum contains nitrogen. The oil of Baku and California are very rich in N, but this element was also found in Pennsylvanian oil, where it had not been found before. The quantity of nitrous products in some crude oil may reach 10–20%. In the oil of Japan, the quantity of nitrogen reaches 1.5%, and in the oil of California it oscillates between 1.00–2.75%. These nitrous compounds are similar everywhere. They are clearly connected with organic decomposition of nitrous bodies of animals and plants, and they cannot originate from the Earth's depths, because they decompose at high temperatures.

---

29. Benzol is a distillation product consisting of benzene, toluene, and the xylenes. Of this mixture, only benzene has the formula C_6H_6, although the other compounds are closely related: toluene has the basic benzene structure and one methyl ($-CH_3$) group attached; xylenes are benzene with two methyl groups. Cyclohexane is C_6H_{12}, and there are other related cyclic hydrocarbons. – *Ed.*

Humans destroyed and are destroying these precious products with an extravagance characteristic of ignorance – without thinking about the future. These nitrous bodies are derivatives of methylquinolene. This discovery is of great significance for understanding the genesis of petroleum.

Methylquinolenes are derivatives of quinolene, the molecule that unites the nucleus of benzene with that of pyridine. Quinolene can be considered as naphthalene, in which one of the CH groups in a location α is replaced by an atom of N. For quinolenes the number of replacements is equal to seven. The derivatives of quinolene and methylquinolene play an important part in the structure of organic compounds of living matter. They can be easily obtained by means of decomposition of different kinds of organic matter, both plant and animal. Quinolenes are the basis of many alkaloids, which are very stable and extremely abundant in terrestrial living matter. The formation of alkaloids in plants is thought to be connected to proteins.

It is impossible to imagine the genesis of the many hundreds of thousands of tons of nitrous bodies in crude oil otherwise than in connection with living matter. These compounds are formed at temperatures that are not very high, and they are quite stable at temperatures slightly exceeding 100–150 °C. Inevitably they will have decomposed at the temperatures of hydrocarbon formation (CO_2 or CO) in the presence of metallic carbides. Besides, their non-organic genesis is incompatible with the geochemical history of nitrogen. The geochemical history of nitrogen resembles the history of carbon, but with great alterations, depending on the gaseous nature of native nitrogen accumulating in great masses on the Earth's surface; an immense part of the free nitrogen is undoubtedly of vadose origin. The free nitrogen corresponding to the carbonic acid in the geochemical history of carbon is the main juvenile mineral of this element. It is stable in all the known envelopes of the Earth's crust.

The Earth's depths possibly contain *metallic nitrides*. Sylvestrite, a nitride of iron, sometimes forms a coating on the lavas of [Mt.] Etna and is a secondary product created by the effect of atmospheric nitrogen on the melted lavas. A. Brin (1905) points at the presence of iron nitrides in lavas and the probable presence of the hypothetical silicon nitride (marignacite). These statements are worthy of attention. The emission of chlorous and fluoric ammonium by volcanoes is without doubt. Only part of them can be related to the decomposition of nitrous remains of living matter overtaken by lava. The emission of ammonia together with overheated water vapors (up to 190 °C) in a geyser area from depths beyond 200 m, such as in Tuscany in Italy or Sonoma in California, cannot be related to the phenomena of life. These gases are of magmatic origin, and they are emitted together with water vapor.

Apparently, ammonia kaolin aluminosilicates are contained in isomorphic admixtures of minerals of volcanic and massive rocks, and the existence of primary nitrogen in these rocks seems very probable. Maybe they contain cyanic and thiocyanogen compounds, as A. Gautier thought. But these compounds are always simple, usually binary. The formation of complex compounds analogous to methylquinolene has never been observed outside the realm of living matter. Only under life's influence does nitrogen combine with carbon, hydrogen, and sulfur. We know another force acting in the same direction, giving simpler nitrogen compounds and manifesting itself in the upper gaseous geospheres of the Earth.

This force is presented by electric discharges and ultraviolet rays. Under their influence, simpler compounds are formed, such as ammonia and oxygen compounds of nitrogen. But this latter reaction is inseparably connected with living matter, as free oxygen is mainly a product of life.

The formation of derivatives of methylquinolenes contained in petroleum corresponds to the geochemical history of nitrogen as we now know. The derivatives of quinolenes are compounds created by organisms that remained unchanged in structure during the metamorphism of the remains of organisms, and passed on to petroleum. The presence of quinolene in crude oil is not only inexplicable, but it also contradicts the most common facts of nitrogen geochemistry if the formation of quinolene is related to magmas.

We arrive at the same conclusion studying other elements of oil such as sulfur, oxygen, and phosphorus. Their compounds are always present in petroleum in different quantities. The quantity of sulfur, for instance, reaches 2.75% in the oil of Texas and Louisiana, and 4 – 5.49% in the Tchussov deposit; part of it is contained there in the form of dissolved free sulfur. The compounds of sulfur present in oil, among which derivatives of thiophene prevail, are apparently diverse, but they are insufficiently studied. The properties and reactions of thiophenes and their derivatives are quite similar to properties and reactions of benzene and the aromatic compounds. The existence of these very stable compounds is also characteristic of such organogenic minerals as coal, whose biogenic nature is no longer doubted.

In this particular case, we again come across a general phenomenon. Throughout nature, molecules with very stable cores of aromatic [ring] molecular structure are widely spread, such as compounds of nitrogen and sulfur in crude oil, and kaolin aluminosilicates and organic compounds in coal. The origin of the sulfur in petroleum, bitumen, and amber is often considered secondary; they are thought to absorb sulfur after their formation, whereas in that state they are easily subject to the action of H_2S, which is widely dispersed in nature. But this hypothesis is absolutely unnecessary, as it cannot be applied to coal, containing, as we have seen, sulfur in the form of analogical cyclic compounds.

The amount of oxygen in oil reaches 6%, and its usual content is 3%. The study of these oxidized compounds, the quantity of which considerably exceeds nitrogen-containing molecules, has long been ignored. They have long been considered secondary products that formed on the Earth's surface under the influence of free oxygen in the atmosphere and vadose waters. This explanation is partly correct, but it can hardly be doubted now that petroleum also contains primary oxidized products in rather large quantity.

Among them, a significant part – up to several percent – is made up of the so-called naphthene acids (V. Markovnikov), ketones (M. Tichvinski), and fatty acids. These are alicyclic compounds with stable rings; some compounds with five-member rings have been found (N. D. Zelinsky, A. Tchitchibabin). The group of these compounds is just beginning to be seriously investigated. Of great interest from the standpoint of oil genesis is the presence of oxygen compounds rotating the plane of polarization of light. According to the hypothesis of Marcusson and Angler, the presence of derivatives such as cholesterin (or phytosterin) explains the rotating capacity of oil.

Cholesterol is a single-base alcohol whose structure has been insufficiently studied, and which plays an important role in biological processes. It forms in plants and concentrates in the bodies of higher animals, receiving only part of it through food.[30] In an egg yolk its quantity exceeds 2%; still more of it is contained in a human brain (2.5%). It can be found in every animal tissue and liquid, and it is also observed in soils. N. D. Zelinsky managed to create a mixture of hydrocarbons with properties of petroleum by means of simple reactions proceeding from cholesterol. These experiments confirm the basis of Marcusson's and Angler's hypothesis. Although cholesterol was not found in petroleum, the products of its transformation are probable. But other oxygen compounds, such as some naphthene acids and hydrocarbons, are also optically active.

Only as late as 1922, the presence of phosphorus was found in Californian oil in considerably larger quantities (0.01% P) than in natural waters. The exact character of the compounds containing P is unknown as yet, but the presence of phosphorus is important as new evidence of the biogenic origin of petroleum. The optical properties of oil provide other proof that their inorganic genesis is impossible. This proof seems irrefutable and clearly indicates their biogenic origin.

All the hydrocarbons produced artificially, and not genetically connected with life products, arc optically inactive.[31] The same is

---

30. It is now known that cholesterol very rarely forms in plants. It forms in animals including humans. There are many related sterols and steroids in plants, but not cholesterol. Readers will recognize that *much* has been learned about the chemistry and physiology of cholesterol since Vernadsky's time. –*Ed.*

31. Traces of optically active molecules (e.g., amino acids) have now been observed in meteorites. –*Ed.*

true for the hydrocarbons that in different cases are proven to form in terrestrial processes outside the biosphere and its organogenic bodies. Now it can be asserted with great confidence that all such bodies are always and inevitably optically inactive. As L. Pasteur has shown, only one medium is known in nature that creates significant carbon molecules in a peculiar manifestation that he called *dissymmetry*: life and its manifestation in the biosphere – living matter.

Dissymmetry, or right- and left-handedness, is one of the main manifestations of life. The symmetry of a natural phenomenon is one of its main properties. Pasteur was a long misunderstood pioneer in this area. Another outstanding French scientist, Pierre Curie, had begun to follow his steps and deepened the understanding of dissymmetry, but death put an end to his studies. Curie was not only a first-class physicist, experimenter and a naturalist with a deep generalizing mind, he was also a great mathematician. He transferred the principles of symmetry to the realm of all physical phenomena; from crystallography where it reigned, he began to create the teaching of the symmetry of physical phenomena. He revealed its utmost logical and empirical significance and connected the idea of symmetry with another major scientific notion that seems clear to us, that of extent. Thus he showed a close relation between symmetry and space in our contemporary understanding of space-time.

The wide understanding of the idea of symmetry by Curie allowed the notion of dissymmetry introduced by Pasteur as a characteristic manifestation of living organisms in their environment to deepen. Pasteur connected the principle of symmetry with the manifestations of right- and left-handedness, which had long drawn the attention of biologists and showed their geometrical manifestation in dissymmetry.[32] Right- and left-handedness in such a statement is connected with enantiomorph manifestations of crystal symmetry of a certain structure. But dissymmetry clearly differs from crystallographic enantiomorph structures in that the equality of right and left manifestation is absent.

According to Curie's principle, dissymmetry is characteristic of life phenomena and can be caused only by something possessing the same kind of dissymmetry. I cannot dwell upon this phenomenon here, I can only draw attention to its importance. I have touched upon it elsewhere. But it is necessary to stress the importance of dissymmetry, which for the first time allows the scientific embrace of right- and left-handedness that is so significant in life phenomena. Furthermore, it allows the problem of oil genesis and its connection with life phenomena to be solved completely.

Living matter consists of chemical molecules with enantiomorph structure in its dissymmetrical manifestation, and it can be a beginning of the formation of new enantiomorph bodies of the same dissymmetrical structure. When living organisms die, they leave dissymmetrical structures in the remains of their matter, in the compounds of their bodies; these structures would otherwise be absent from the Earth's crust, and probably also from the planet as a whole. An indisputable statement can be made that *in the Earth's crust, only life and the material products of its decomposition can*

---

32. Strange as it is, the ideas of Curie and the idea of symmetry itself have not entered the scientific mind to a complete measure. His deep substantiation of dissymmetry has no influence upon science as expected; his idea has not been developed philosophically. [The importance of the dissymetry of the molecules of life – e.g., sugars and amino acids – is now taken for granted and understood in considerable detail. –*Ed.*]

possess dissymmetry; i.e., the potential of manifesting right- and left-handedness unequally.

For living matter this property manifests itself also in another phenomenon, which makes it as different from the inert bodies of the Earth's crust as the dissymmetry of the molecules that build up an organism. This is the regeneration of a living organism – the proliferation and incessant creation of dissymmetrical bodies. We know that living matter has existed for millions of years without a break, and that in natural phenomena its spontaneous origin is impossible. The principle "Omne vivo e vivo" is completely within the framework of the empirical generalization substantiated by P. Curie: dissymmetry can be engendered only by a similar dissymmetry.

Life does not exist in the Earth's depths where the genesis of oil had been looked for. In order to explain the formation of enantiomorph structures like the crude oil there, it is necessary to allow the existence of a similar dissymmetrical environment at these depths. Our present-day scientific knowledge gives no indications of this. Remaining on the grounds of exact knowledge, we must admit that the phenomenon of the optical activity of a carbonaceous mineral, as it is related to the inequality of its right-handed and left-handed manifestations, inevitably leads to living matter for molecular structures containing carbon atoms – the only physical medium in which dissymmetry exists. It is absolutely incomprehensible that the optical activity of crude oil has so long been ignored and that during the creation of the numerous theories of their genesis it was not taken into account.

The French physicist Biot discovered in 1835 optical activity of oil of unknown origin. He paid no heed to this discovery and mentioned

oil incidentally, amongst many other bodies in his work. He systematically and experimentally stated the presence or absence of the
optical rotation of [polarized] light in a series of natural and artificial
compounds. This important observation was completely forgotten.
True, the sample of oil used by Biot was of unknown origin, and,
as we know now, had unusual properties (left rotation). Besides,
Academician E. Lenz, studying the Baku crude oil especially for this
purpose, came to a negative conclusion in 1833.

The rotation of polarized light by crude oil was noticed as early as
the nineteenth century. Almost simultaneously (around 1900), L.
A. Chugaev in Moscow, and P. Walden in Riga, paid attention to the
existing data in the literature about this rotation. The first author
noted an indication in the technical literature about the rotation of
oil products such as Vaseline oil, which he checked experimentally;
and the second mentioned the importance of Biot's discovery for
understanding the genesis of crude oil in a few lines of his article
about the development of stereochemistry. But only as late as 1904
were these forgotten and half-forgotten facts revived.

In January, at a session of the Naturalist Society in Moscow, and in
connection with the new theory regarding crude oil bedding put
forward by the original and talented scientist A. P . Ivanov (1865–
1933), an argument about the genesis of oil began, which made this
phenomenon clear. L. A. Chugaev undertook deep and broad research
into the issue of dissymmetry of life products in connection with oil
genesis. M. A. Rakuzin, who was present at the session, decided to
check this phenomenon experimentally, and soon, using the samples
of Baku, Grozny, North American petroleum and processed products,
proved the *right-handed* optical rotation of oil in the studied cases.
For a very long time, all the crude oil that was being studied had

shown right-handed rotation, and only several years after the studies of P. Walden, some oil with left-handed rotation was found, but it was much weaker than that studied by Biot.

The facts remain firmly stated: petroleums are bodies of an optically active structure with a distinct difference between the right and the left antipodes. Right-handed oil clearly dominates in nature. Such carbon compounds are created only in the biosphere and only by living matter. All carbon minerals without biochemical genesis are optically inert. Now it can be asserted that this property is characteristic not only of oil, but of all the remains of organisms' bodies; different compounds with a distinct dissymmetry can always be discovered in them. The matter of the biosphere is not physically uniform; living organisms and organogenic carbonaceous solid and liquid minerals make up a dissymmetrical medium, which is incorporated into the inert matter that lacks it.

The dissymmetry of coal and the humus of the soils follow from the crystallographic research of L. Royer, although he had not arrived at this conclusion himself. Royer showed the dissymmetry of crude oil in its new manifestation: In the presence of oil, the figures etched on the crystals that possess a center and symmetry planes are enantiomorph, and in the studied cases right-handed. The same was proven by Royer for soil humus and for peat bogs. A vast area untouched by scientific study is opened before us.

The chemical study of petroleum thus leads us to conclude their biogenic origin and reveals the fact that juvenile hydrocarbons cannot play a significant part in them. The study of oil by geologists and biologists leads to the same conclusion. Crude oil is not vadose mineral; it does not form as a result of weathering in the presence of oxygen. it is *phreatic* or deep vadose mineral,[33] and the temperature at the time of its formation is not very high. But its initial matter cannot originate from the deep areas of the Earth's crust; it must be looked for in the living matter of the biosphere and under special conditions.

There is still much to be understood about the origin of petroleum, but many aspects are becoming clear and we are on the right path. First we have seen that crude oils are not only hydrocarbons; besides hydrocarbons, they include other complex organic compounds with clear signs of biogenic origin. Unfortunately, we are still completely lacking rationally conducted analyses of the composition of such oils. What quantity of hydrocarbons do crude oils contain? This quantity is probably variable, and at present it can only be assumed. The average is apparently somewhere between 75–95%.

Petroleums form in the biosphere. They are products of the metamorphism of decomposed and (biochemically) decomposing remains of plants and animals. The metamorphism itself is probably a biochemical (maybe a radiochemical) process; the usual limits of the biosphere (the lower limit) should be revised. But where does the genesis of crude oil begin on the Earth's surface, a beginning that must manifest itself as certain biocoenoses or accumulations of detritus material, the remains of dead organisms?

For a long time this had not been known. Only more or less probable assumptions had been made, to which the authors themselves had not paid much attention. Many theories can be found in the history of

---

33. Maybe it will be possible to speak about the vadose origin of oil if this turns out to be a biochemical process. I use the term "vadose" here as applied to the biosphere.

natural science that approach the present-day scientific statements, as for instance those put forward in 1850–1860 by the well-known geologists J. S. Newberry (an American) and H. Abich (a Russian-German scientist). Only in this century was it possible to discover the everyday phenomena that had long seemed insignificant, but that, in reality, engender the marvelous phenomenon of petroleum genesis.

To understand them, hard and profound work was required. New sciences appeared such as marsh study, plant ecology, and limnology; the study of peats and silts got a new direction. New, large branches of science have been created with their own methods and problems, such as the chemistry of coal and crude oil, or mineralogy of carbon and caustobiolytes. The position of crude oil among other organogenic carbonaceous compounds, and the necessity of studying them simultaneously as a result of the geochemical work of life, have become clear. Only at present, with the appearance of geochemistry and its peculiar branch biogeochemistry, the importance of this direction of scientific thought has received a rational basis.

In order to discover the genesis of petroleum, it is necessary to study the organogenic products of the death of certain organisms, as well as the biochemical processes of their transformation by other organisms both in the biosphere and at the transition to the next terrestrial envelope – the stratisphere.[34] The formation of oil is an extremely important manifestation of the process in which solar energy is transferred to deeper layers of the planet through living matter.

Here, several different chemical processes of such transfer can be mentioned, one of which is oil genesis. For example, the creation of dispersed organogenic matter, or the creation of compounds with a

---

34. That is, sedimentary rocks; strata. –*Ed.*

prevalence of hydrocarbons (oil), compounds rich in hydrocarbons (bitumen), or poor in hydrocarbons (coal), and almost pure carbon (graphites). The initial matter for the formation of crude oil can be accumulated in the biosphere in two ways, and they should be distinguished: the accumulation of dead organisms in the area where the organisms lived or in the areas to which their remains were carried.

Examples of the first kind have been stated for fresh and salt lake basins, and for areas of rivers where the remains of dead organisms accumulate, or where the current is very slow. But such accumulations can sometimes, under special conditions, form in the seas. Thus, the formation of the oil of the Northern Caucasus and the Kertch Peninsula is possible in connection with a peculiar phenomenon, such as a powerful development in the bottom and silt water at the expense of the main sea water. This is still going on in the Black Sea in connection with the emission of hydrogen sulfide and which, according to A. D. Archangelsky, also took place in the Miocene and Oligocene.

There are similar conditions known for petroleum formation that involve diatom plankton, which is accepted for Californian oil and considered impossible for the Baku area. In certain cases, crude oil formation can be connected to the blooms of organisms in the sea, such as for instance the blooming that is observed at present in the shallow and salty Adriatic Sea – the "mare sporco" [red tide]. Direct determinations of the presence of mineral oil in the sea and ocean silts have given negative results, as could be expected. Now only general remarks can be made about where oil formation begins in the biosphere, but the direction of further work is clear. It could be discovered only after the processes that go on around us, and seem trivial and insignificant, such as the processes of decomposition or decay of organisms, have become the object of thorough scientific study. The process of coal formation has been revealed partly in this way. Crude oil and coal are different members of the same natural process, requiring decomposition of dead organisms in water without access to free oxygen, and the concentration of such decomposed elements in large masses.

Of primary importance was the gradually growing belief that the carbonaceous minerals created in this way differ not only in their dependence on the external conditions of decomposition, but also in chemical composition regarding the organisms they originated from. Of great importance were for example the work of the French paleophytologist, B. Renault, and his pupil, C. Bertrand, undertaken at the turn of the nineteenth century and not understood by his contemporaries. They discovered the formation of bituminous coal and proved their difference from other coal, both in respect to the organisms (unicellular algae and small animals) from which they had originated, and in respect to their chemical composition and structure. They saw how this process involved the formation of hydrocarbons.

The German paleophytologist, H. Potonie, having turned to observation of present-day nature and processes of decomposition of dead organisms under water, discovered the diversity of accumulations and their dependence on certain organisms. At the same time he pointed out that in fresh water silt accumulations, products concentrate (sapropels and sapropelites[35]) that are distinctly differ-

---

35. Vernadsky uses a number of terms that are not familiar to those of us who don't work primarily in the field of geology. *Caustobiolyte* does not occur in any of my dictionaries. It apparently refers to minerals (in the broad sense) that had their origin in biological materials. *Sapropel* is mud consisting chiefly of decomposed organic matter formed at the bottom of a stagnant sea or lake, and *saprolite* is a soft, disintegrated, usually more or less decomposed rock remaining in its original place. –*Ed.*

ent from ordinary coal, and are formed mainly from animal organisms and plankton species that are rich in hydrocarbons. Coal is connected with arboreal plants.

It must be borne in mind that these dead masses of life are filled with life – bacterial and maybe fungal – when they are in the biosphere before the metamorphic processes begin. According to Waxman, their mass includes more than 30% of living bacteria. H. Potonie and other scientists have created a classification of natural decomposition based on the products of organisms' remains, caustobiolytes as they are called. Leaving out all the particularities that have no importance in our case, we can point out three distinctly different types of such change:

1. Unicellular algae and microscopic animals: As a result, bituminous shales appear that are rich in hydrocarbons.
2. Decomposition products of green marsh mosses and grasses: peats appear as a result.
3. Decomposition products of forests and arboreous marsh plants: coal is created.

No doubt, in all these cases only the dominant process can be mentioned. The study of present-day processes in water reservoirs involving transformed remains allows a detailed reconstruction of the past. In general, it turns out that all these organogenic carbonaceous bodies, as far as their accumulations are concerned, appear in definite areas of the Earth's crust where life is particularly concentrated: in coastal concentrations on the border of sea and land and, to a much smaller extent, in the fresh or salty accumulations of lakes. The importance of coastal concentrations indicates both the possibility of their formation in separate areas in the sea medium and their connection with orogenic processes by which the character of the coastlines is determined.

This concentration of life is determined by the main process of terrestrial decomposition on land, erosion by rivers and meteorological water, the sea surf, and the drift of a great quantity of nutritious matter for living organisms in a form that can be well assimilated. These areas, in case the erosion of the land provides a lot of nutritious matter, present convenient conditions for preservation and decay of organic matter, and are the biospheric origin of both crude oil and other organogenic minerals.

The processes of decomposition giving peat, resinous coal, crude oil or coal, occur in a water medium poor in oxygen or containing no oxygen at all. They begin in the biosphere or in the areas close to it (the upper stratisphere). It is becoming clear that these minerals are formed biochemically. A great role in the creation of natural carbonaceous compounds is played by anaerobic life – bacteria that have penetrated deeper into the Earth's crust than we previously thought. They reach areas lying far below the oxygen surface, beyond the crust of weathering, of steady, unchangeable temperature exceeding the average temperature of the air. The forgotten observations of F. Stapf (1879) that were found recently (November 1926), are an unexpected confirmation – thanks to the discovery by E. Bastin and N. Ushinsky – of bacteria living in waters of oil-bearing stratum at depths of more than a kilometer, such as in North America and on the Apsheron peninsula in the Caspian Sea near Baku.

The area of the biosphere must be expanded far into the depths, far beyond the oxygen surface. Anaerobic life should be taken into account, which may be much more developed in the waters of deep strata than we are currently thinking. Another phenomenon seems to bring us to the same conclusion about the expansion of micro-flora to the depths in a latent state. The organisms' remains, which will eventually turn into oil, move to a depth of a kilometer and deeper due to geological shifts; they may remain in this area of life for a long time and undergo biochemical transformations.

Remains get to these depths usually as a result of geological shifts, for instance, when still waters in which decomposing living matter has accumulated become covered with precipitations, and in this way, or by the movements of the Earth's crust, pass on to its deeper layers, beyond the limits of the crust of weathering. Vertical sections of peat bogs and sapropels clearly show that chemical transforma-tion takes place in their deep lower layers, which become denser and turn into peats or sapropelites (bituminous slates), containing the chemically transformed life products.

The formation of final products takes place in a geologically insig-nificant amount of time under favorable conditions. The signs of quick formation, such as the presence of shells whose bodies have turned into oil, or coal boulders in the intermediate layers of coal deposits, presented a difficulty for scientists at first. They become incomprehensible if the biochemical origin of all these bodies and the participation of anaerobic bacteria are not admitted. At present we are only beginning the microbiological study of caustobiolytes' formation. Evidently, our knowledge will soon become more exact and clear. But even now, the inevitable acknowledgment of the anaerobes' participation indicates clear limits to the area in which the process of oil formation may take place.

It cannot take place beyond the limits of life's existence. The maximum possible temperature for this process hardly exceeds 60–70 ºC, and was probably lower still. A higher temperature must lead to the formation of gaseous products and to the disappearance of carbonaceous compounds in the given area. Thus, taking into account the thermal gradient of the Earth's crust, the formation of oil could probably not take place deeper than 1.5 km. It is likely to have occurred within the limits of a few hundred meters. Furthermore, the pressure could not have exceeded several dozens of atmospheres.

Biochemical decomposition of organisms' remains must and does go on in different ways for different organisms. Organisms that are dispersed amongst other, alien organisms, decompose and yield their own products characteristic of their species and their chemical composition. The appearance of different kinds of caustobiolytes such as *crude oil, coal, and bituminous slate* results from a considerable prevalence of organisms with a certain chemical composition in each case. Here we see a manifestation of an often forgotten phenomenon – the great chemical variety in the composition of living organisms.

For oil and other caustobiolytes the main requirements must be met: the formation of large accumulations of organisms of a certain chemical composition on the Earth's surface, and the creation of certain conditions that are favorable for their anoxic underwater decomposition. All we know about the accumulations of caustobiolytes shows that their formation mainly took place where the *organisms* had lived. In the vast majority of cases, not the organisms themselves had shifted, but the *region* of their habitation.

This is also true for liquid products such as petroleum. The majority of large oil deposits correspond to the region of their origin and creation. The so-called "migration" of oil due to its easy mobility is in general a secondary phenomenon. It can play a role in creating reservoirs of oil after their formation, but not in the creation of the oils themselves. This does not contradict the fact that the presence of oil is closely connected with dislocations of the Earth's crust, and concentrated mainly in tectonically damaged areas. Deposits of coal also concentrate mainly in such areas. This is not related to the mobility of oil, but to the existence of favorable conditions (a favorable regime for the surface and stratum waters) for the creation of caustobiolytes, and for the preservation of the organisms' remains in these areas.

In such areas subject to tectonic shifts (for instance connected with geosynclinal shifts or those along the edges of these areas), favorable conditions appear for the creation of reservoirs (fresh- and salt-water reservoirs at the borders of sea reservoirs), and their subsidence, due to which accumulations of life appear and powerful layers of caustobiolytes form. But in these circumstances, conditions are created for further shifts and divisions of the formed mixture of solid, liquid, and gaseous bodies of "primary" crude oil. Sometimes these conditioned divisions begin to bring about the corresponding shifts soon after the formation of oil. Thus, concentrations of liquid rich in carbon bodies appear in the same way as is observed for other liquids of the Earth's crust, such as water solutions.

These processes depend on the conditions of the environment, the geological structure, tectonic shifts, properties of oil, etc. They lead to the formation of oil deposits. Here I will not touch upon this important branch of geology in which much is still unclear, but which will surely be embraced by science. I must only mention that all these phenomena take place after the formation of oil – this certain caustobiolyte that is closely connected with living matter. The enormous geological force of this substance is clearly seen here. But as we shall see below, the power of its influence goes much further, because only a part of the organism's matter accumulates in the form of caustobiolytes (a hundred-thousandth of a millionth part of the chemical elements passing through living matter). The remaining mass of the elements is held within the cycle of life, in the realm of living matter's activity.

This tiny part, however, manifests itself in the vast masses of oil or coal, and in the hundreds of millions of tons of carbon. It can create these masses only under special geological conditions, those that allow it to leave the realm of active living matter in a geologically short time; i.e., leave the biosphere, and in particular the crust of weathering. Such conditions exist only in areas of great tectonic shifts. The formation of oil or coal is determined by the character of the accumulation of living matter (organisms) that has appeared in the biosphere on land, and that has then been removed from the upper zone of the biosphere for geological reasons.

Crude oil and coal can form only in definite types of life accumulation in the biosphere: in fresh water (less frequent in brackish and salty) accumulations on land. The chemical composition of the organisms of the accumulations determines the genesis of either coal or oil. As a result of this process, not only physically different products – solid coal and liquid oil – appear, but also their chemically different mixtures. At present, understanding of the chemistry of coal is making rapid progress, and we see a picture that is distinctly different from the one we have just shown for the chemistry of oil. Hydrocarbons play an insignificant role in coal, and more complex humic oxygen- and hydrogen-derivatives of carbon are present in the foreground.

The chemical structure of these humic parts is becoming clear at present. Schrader and Fischer stressed the importance of the lignin group – the principal part of timber – that is preserved practically unchanged during formation of coal. Lignins are numerous; the formulas of their molecules cannot be given exactly and definitely, but their main structure can be approximately stated. The composition of lignins is expressed (Fuchs, Stadnikov) by the percentages:

C: 62–69%, H: 4.5–6.6%, O: 25–32%

A characteristic feature of lignins is a nucleus of aromatic character, and the coniferin complex prevails. The chemical function of the very stable and inert lignins is not clear. It always contains metaxil groups (CH_3O) up to 13.03%, explaining the formation of alcohol from wood. Apparently, the number of metaxil groups in the coal of the Carboniferous is smaller than in that of the Oligocene. The number of lignin kernels in coal changes as it changes in plants.

Lignin is contained in all plants except for algae (and unstudied microscopic plants). In peat mosses, its content oscillates between 9–13%, and in tree-sized plants between 20–30%. It seems to be one of the most widespread materials created by life in the biosphere. In coal, the content of its derivatives also varies, probably depending on the composition of the organisms they originate from. Anthracites and bituminous coal are poor in derivatives, and coking coals are very rich. During the process of metamorphism, the derivatives of lignin in coal condense and become enriched in carbon.

Its dominance over the derivatives of cellulose in carbonaceous minerals originating from living plant matter is important. On the scale of the Earth's crust, a process takes place that is characteristic of the organism's metabolism: the creation of compounds with a stable *annular* nucleus in the structure of molecules. If we knew the chemistry of organisms and their different species as well as we know the chemistry of oil and coal, we could undoubtedly distinguish types of organisms as different from each other chemically as they are. Unfortunately, we are only just beginning to pay attention to these areas of biology.

The development of our notions about the genesis of caustobiolytes is not yet finished, and its main features mentioned here are not yet generally acknowledged. However, the study of the history of thought in this field leads to the same conclusions as the direct study of facts, which allows for more confidence in expecting the confirmation of the correctness of the ideas presented here. A decisive impetus to scientific thought was given by the work of B. Renault and C. Bertrand, which proved the differences in biogenic origin for natural coal. The influence of these researchers on the extensive work of numerous scientists studying these matters so important for civilization was very great.

Throughout generations, work continued unsystematically; the particular history of separate coal deposits was studied in detail, and no generalizations were made. A phenomenon was usually not compared with the contemporary natural world. The biogenic origin of crude oil was first embraced by the thought of Buffon in the middle of the eighteenth century and became commonly understood as late as at the end of the nineteenth century. Regarding coal, as early as in the sixteenth century, V. Cordus (1544) and B. Klein (1592) understood their origin as being tree-sized plants. By the end of the eighteenth century this had become a common idea; through Hutton (1795) and the synthesis of the Earth's history connected with his name, it had entered the geological mind. Paleophytologists of the beginning of the nineteenth century (Sterner, Brogniart, etc.) laid exact foundations for our present-day understanding.

This present-day understanding, however, is not based only on the data of paleophytology. Its roots are also in chemical and geological research on coal. Apparently, the contemporary ideas had first appeared in a more or less clear form in the works of Belgian geologists. They were expressed by the Belgian mining engineer A. Briard, who worked in the mines of his country's basin, and who developed the ideas of A. Bougui (1855). But his work attracted no attention and could not influence science. Only much later, the American scientists L. Orton and J. S. Newberry came to ideas close to the ones of our days. There were many outstanding scientists among the geologists and paleophytologists. I cannot enumerate all of them, but I shall mention the works of H. Filhol, C. Grand' Eury, and M. D. Zalesski. All of them were brought to light by the discoveries of B. Renault and C. Bertrand.

The formation of coal is connected with marshes, with large accumulations of plants characteristic of countries with moist climates, for example in the mouths and deltas of big rivers, in the plains of their basins, on the coasts of continents and islands, and in the lowlands of tidal areas. All these are accumulations of life in which a large mass of organic matter is present in a state of slow decomposition. Possibly, they are the largest terrestrial accumulations of life we know.

Marsh grass and mosses cannot be the origin of coal and crude oil. Lignites and coal originate from the marshy forests of subtropical lowlands, or the marshy plains of the river basins in different climatic belts. Perhaps under favorable geological conditions, the marshy forests of Florida could become strata of coal after many millennia.

The present-day abundance of forests in America is partly related to these marshy forests, which have masses of tree-sized plants per hectare unknown in the forests of our latitudes. Possible sources of coal (if humans did not prevent it) could appear also in other areas in the U.S.A, like the Dismal Marsh in the states of Virginia and North Carolina, or lake coasts like the Upper Klamath in Oregon.

Another physical and geographical phenomenon must furthermore be taken into account: the accumulation of life in the basins of great virgin rivers. The Amazon, Orinoco, Zambezi, Ob, Irtysh, and their tributaries are some examples, having forests on their banks, vast flood plains, and marshes. These rivers with their water rich in dissolved organic matter and plant remains, with the lakes through which they flow slowly, and with their mouths and deltas, give us an idea about the initial formation of a coal basin. "The Vasugan Sea" of the Ob stretches out for several thousands of square kilometers.[36] There are other large areas of low coastal land covered with vast marshy forests, such as flood areas in tropical and subtropical zones, both on the main continents and on islands. In India and on the islands of Indonesia we see examples of these phenomena at present. The coal strata in England, which constitute its main wealth, are metamorphic remains of forests of such big river deltas.

Both coal and lignites, as well as crude oil and resinous shales, are products of changes going on beyond the green surface, in the anaerobic medium. Apparently, the water solution acts here first of all. This is pointed out by the observations of Taylor who has paid attention to the differences of rocks covering coal deposits. When

---

36. Now it is a tremendous and widely exploited oil and gas basin. [*Comment by the editor of the third edition.*]

they are covered with strata creating an acid medium, peats form; when the medium is rich in Ca, Al, and Si, lignites form; and when it is rich in Na, Al, and Si (alkaline medium), coals form. This is the first important generalization now being tested.

The formation of bog heads, bituminous slates, and crude oil happens in a different way. It also begins in lakes and other still waters, or coastal parts of the sea that are shallow, fresh or salty, and rich in living matter. The *living matter* there is very different from that prevailing in the water reservoirs where peats or coal form. This refers to the organisms of plankton, unicellular algae, small animals, and spores. They accumulate thanks to explosions of proliferation similar to the "blooming" of our lakes and pools, to the blooming of seas. This occurs on the surface of the water reservoirs under the influence of "the waves of life" and the great geochemical energy of these organisms. In a short time these surfaces become covered with myriads of organisms: dozens or hundreds of thousands of unicellular algae and other protophites that are usually of the same species. Dozens of species of organisms are known which at present give such "blooms" in surface waters.

The rich phytoplankton of these still waters brings about an unusual increase of aquatic animal life. Such "waves of life" disappear as quickly as they appear. Organisms die without being eaten up by others, and their remains accumulate on the bottom. A large part of their matter leaves the cycle of life in this way, and becomes sapropels, rich in oily and nitrous compounds, closely connected with oil. At present, sapropels also accumulate in thousands of lakes in the Northern Hemisphere, which are morphologically related to the last glacial cover. This occurs for example in thousands of lakes of

Eastern Siberia and in all the places where conditions are favorable for these organisms to flourish.

In the tropical black rivers, which are sometimes almost still, conditions are often very favorable for the slow decomposition of dead organisms, as these waters rich in dissolved organic matter are poor in aerobic microbes (like in peat). Their organic matter slowly turns into gaseous products, but they accumulate some of the remains of life taken there. Sometimes, animal remains play a minor part, and the waters contain spores brought in by wind, or nitrogen-concentrating fungi.

These two different processes of living matter decomposition are not always separated from each other. On the contrary, they almost always take place simultaneously. If organisms rich in derivatives of lignin dominate, the conditions are favorable for formation of coal. If organisms that form sapropels dominate, bitumen and oil will follow. The accumulation of living matter is noticed only when processes take place on a large scale. But eventually it is impossible to ignore the transformation of disparate remains, insignificant in itself, which takes place around us incessantly, since the Earth's crust possesses mechanisms allowing the insignificant dispersed masses of these products to accumulate and make up larger masses. These are mainly processes taking place outside the biosphere, in the stratisphere [sedimentary rocks], when the organogenic remains of life leave the limits of the Earth's surface.

This is the process of slow (first biochemical) metamorphism, which eventually gives pure carbon and carbon oxide. But before passing to this final state, coal and bitumen have a very long history, existing for millions of years in a state of slow transformation. Geologists who have studied large oil deposits have gradually come to analogical ideas. Two large scientific institutions, The U.S. Geological Survey in Washington and the Geological Committee in this country, had come to almost similar views by the beginning of the twentieth century. This happened after research that lasted dozens of years; after a deep and independent study of the largest petroleum deposits. Oils are undoubtedly sedimentary, phreatic, and aquatic (mainly of fresh-water genesis, but as usual connected with the activity of sea minerals).

Crude oils are situated in certain geological formations of sandstones, sands, limestones, and clays. Their large conglomerates of millions of tons are known from the Silurian to the Pliocene. In all their deposits, either rocks rich in remains of organisms are found, or other indications of the former existence of large accumulations of life that have given birth to them. Geological processes often carry oils away from the locations where organic matter accumulated. Thus, although the very mechanism of the process is not well known, and the formation of oil is to a great extent an arguable point, the main fact has been elucidated: crude oils originate from certain living organisms, from living matter of a certain chemical composition determining the chemical structure of oil. They consist of phreatic compounds whose genesis has its beginning on the Earth's surface; oils are not juvenile minerals.[37]

---

37. In spite of Vernadsky's compelling arguments of 1933 or earlier, there are still those (notably Thomas Gold at Cornell University) who contend that petroleums are juvenile minerals. It has been suggested that they or their precursors were synthesized in outer space and condensed with other matter when the Earth was formed and that they have arrived on Earth at all times during its history from comets and asteroids, which are known to contain hydrocarbons. Evidence for this idea is that oil reserves often occur thousands of meters *below* fossil-containing strata. Also, there seems to be too much petroleum compared with the quantity of living matter that might have been fossilized. Optical activity of petroleum components was perhaps

6 the primary geochemical carbon cycle

The most stable mineral of carbon in the Earth's crust is, without question, native carbon. But it is possible that in the deeper envelopes, such as the basalt one, this native carbon becomes unstable and partly turns into metallic carbides, and possibly into carbon oxide. The formation of native juvenile carbon shows that for this element, the capacity of chemical activity decreases in thermodynamic envelopes with the increase of pressure and temperature. This capacity becomes the more intense the closer it is to the Earth's surface. That is why the quantity of its compounds increases in the upper envelopes and reaches its maximum in the biosphere, where under the influence of the Sun's energy on living matter, millions of its compounds form. Their transformation into 'native carbon' or graphite, a slow, geological phenomenon of long duration, is the most characteristic fact of the primary geochemical carbon cycle.

The biosphere is an envelope of life – the realm of all existing living matter. All life's carbon is captured by the biosphere. All the carbonaceous compounds that exist and form in it are connected with living matter in one way or another. All the phreatic carbonaceous minerals getting there as a result of geological processes basically originate from living matter, and are metamorphic products of vadose minerals formerly connected with life. CO_2 is the only juvenile and phreatic carbon mineral entering the biosphere in great quantities. The purely juvenile origin of carbonic acid can be stated only for some of its

---

Vernadsky's strongest argument, but some organic materials arriving in meteorites are optically active. (A search of the Internet under "oil from asteroids" finds many sites.) –E*d*.

mass. Its larger portion, even in deeper areas, originates from carbonates. Carbonates are almost always vadose or phreatic minerals. Only rarely, under specific conditions, do they separate themselves from igneous sediments. It is important to mention that there are many chemical processes on the Earth's surface connected with the synthesis of carbonic acid. These processes are obviously related to living matter, for all of them occur under the influence of free oxygen. Free oxygen oxidizes carbonaceous and even graphite dust, and large quantities of carbonic acid form within living matter itself, under the influence of the respiration processes.

Hydrocarbons (mainly methane), which no doubt originate from the deeper layers of the Earth's surface, are only partly of juvenile origin. The bulk of their mass forms in vadose areas, such as swamp gases (a biochemical product). The other part is created in the stratisphere, for instance from gases in coal mines. But this explanation is hardly applicable to all the gas torrents of hydrocarbons, a great part of which is incessantly emitted at present by drilling, and which have been emitted under natural conditions over the years. Some of them are genetically related to the gaseous phase of crude oil deposits. Another amount must be brought into accordance with the dispersed organic matter of sedimentary rocks; that is, has a complex origin expressed by the sequence:

Sea life $\rightarrow$ sea silt $\rightarrow$ sedimentary rocks $\rightarrow$ gases

table 8

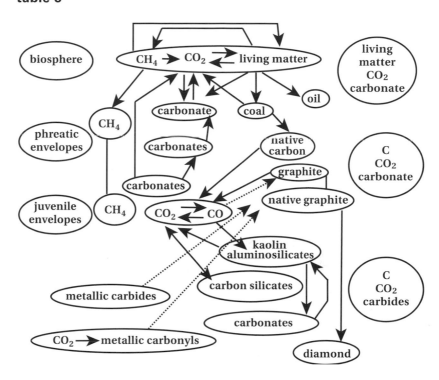

The transformation into gases must take place by means of biochemical and abiotic changes in a medium devoid of oxygen. But still, part of the methane can be connected with magmatic centers, and comprises the deep underground atmospheres of H_2O and CH_4. The genesis of these atmospheres must be complex, and the water vapors and hydrocarbons can be of different origins.

7   earth's gases and living matter

The most important fact in the history of carbon is living matter producing gaseous products, mainly CO_2. Methane and carbon oxide formed by living matter are very rare in the biosphere, as they reach it only as traces and then turn into CO_2. The close genetic connection between the Earth's gases and living matter is not restricted to the creation of carbon dioxide. *Almost all the matter of organisms is created from gases; all the Earth's gases (except volcanic emanations) are in some way connected with life processes of organisms.* At the same time, the same gases: O_2, CO_2, H_2O, NH_3, H_2S, SO_2, H_2, CH_4, CO, CHOH, CSO, and NO_2, are released back into the atmosphere during decomposition upon the organisms' death. No doubt, more than 97–98% of all living matter's atomic mass is produced from the gases of the biosphere, and probably a smaller quantity, but of the same order, is released as gases after the death of organisms. Some numerical data will help us to evaluate this phenomenon better.

A large amount of Earth's matter is annually put into motion by life; this mass is much larger and more significant than we imagine. The Earth's crust weighs about 2×10^{19} t. The mass of living matter is no more than 0.1% of this weight, and corresponds to about 10^{14}–10^{15} t. Much more than nine-tenths of this mass originates from the Earth's gases. Organisms do an enormous amount of work by bringing about this gas cycle that lasts for billions of years. This work is still greater taking into account the fact that in order to sustain their existence, organisms annually move masses of gases that exceed the weight of the atmosphere – 6×10^{13} t – by several times.

In the outstanding synthesis of the chemical status of living matter given by J. B. Dumas and J. Boussingault 90 years ago, they correctly characterized the realm of organisms as "an appendix to the atmosphere." They were right to say so because the incessant exchanges going on between living matter and the Earth's atmosphere are commensurable. "Summing up what we have said," Dumas says while finishing his lectures, "we see that the primary atmosphere of the Earth has been subdivided into three large parts: one of them makes up the present-day atmosphere, the second part is represented by plants, and the third by animals. . . . Thus, everything that the air gives to plants, the plants yield to animals, and the animals return to the air – it is an eternal circle in which life throbs and manifests itself, but in which matter does nothing but change place. The inert matter of the air, which gradually gets organized in plants, begins finally to function in animals and becomes an instrument of thought. Then, defeated by this effort as if broken, it returns as inert matter to the great reservoir from which it has come."

Although in the twentieth century we would alter some ideas that are put forward here, the main statement remains unchanged. Living matter is inseparably connected with the atmosphere; it turns atmospheric gases into colloidal bodies, both solid and liquid. Living matter is not only an appendix of the atmosphere, it is much more significant for the Earth; the atmosphere of our planet taken in mass, not in volume, is probably smaller than that of the living matter that mainly originates from it. Moreover, a deeper study of the phenomenon shows that the influence of living matter on the migration of atoms in the Earth's crust – its geochemical energy affecting the migration – can manifest itself only in close connection with the atmosphere. Finally, the empirical analysis of biogeochemical processes, freeing itself from the delusion of cosmogonies, leads to the conclusion that the Earth's atmosphere itself, consisting primarily of oxygen, nitrogen, and carbon, is the creation of life.

8 living matter and the geochemical history of carbon

The importance of living matter in the biosphere goes far beyond the limits of the geochemical history of carbon. Although carbon is necessary for the creation of living matter, the latter does not become a carbonaceous body due to that. Its function in the Earth's crust is more complex, and its properties are more than the sum of the properties of the organic compounds of which it consists.

The idea, so widely spread a few years ago, that the phenomena of life can be explained by the existence of complex carbonaceous compounds, *living proteins*, has been rejected completely because of empirical geochemical facts. Neither proteins, nor other carbonaceous compounds, protoplasm, or their natural mixture, can give an idea of living matter. *Living matter is the sum total of all organisms.* Its activities are the result of all its matter as a whole. Saying that manifestations of organisms concentrate first of all in proteins, and not in the carbonates of the free atmospheric oxygen they produce, contradicts reality.[38]

In fact, the influence of living matter impacts the entire chemistry of the Earth's crust and directs the geochemical history of almost all its elements. It is obvious that in the geochemical history of carbon,

---

38. Although Vernadsky is technically correct – the manifestations of life can not be ascribed to any single compound – he seems to be unaware of the critical role in life functions of the *enzymes*, all of which are proteins. As a first approximation, it is valid to say that all life function is a matter of chemical reactions controlled by enzymes. We then would go on to describe the role of the nucleic acids in controlling enzyme (protein) synthesis and in carrying the "information" of life from generation to generation as cells divide and as organisms reproduce. The importance of enzymes was beginning to be understood in the late 1800s, and Vernadsky seems to be reacting against some of the statements that were being made about the role of proteins. One gets the impression here and in other parts of the text that, of the myriad scientific fields that Vernadsky had mastered, he was least at home in the field of biochemistry. –*Ed.*

only some problems are revealed before us, giving no full notion of the enormous significance of living nature in the history of our planet. But the geochemistry of carbon cannot be understood aside from the phenomena of life. I will shortly touch upon four life phenomena of great geochemical importance:

1. The spreading of living matter over the Earth's surface;
2. Its mass and chemical composition;
3. Its relation to carbonic acid [CO_2];
4. The role of living matter in the history of chemical elements.

9 spreading of living matter in the biosphere and the geochemical importance of the proliferation of organisms

The spreading of living matter on the Earth's surface is one of its most surprising and exclusive properties. It is a sort of reflection of its gaseous origin. We see nothing like this phenomenon in inert matter, except in its gaseous state. This analogy is logically correct; in both cases we study statistical phenomena, and the laws of gases and the proliferation of organisms have much in common.

We imagine the spreading of gaseous masses and their resiliency as the result of chaotic spontaneous movements of their molecules, which can only be stopped by an external force exceeding the combined force of movement of all these small bodies. Had this external force not existed, gases would eventually have filled all the given space. The same phenomenon is observed in nature, in the totality of organisms. Had no insurmountable obstacles existed, living matter would in time have come to fill all available space through the spontaneous movements of its indivisible "molecules."

If obstacles exist, living matter produces pressure analogous to the pressure of gases.

This pressure has a characteristic feature related to a special state of the indivisible entities constituting living matter, which are only partly analogous to gas molecules. The pressure of living matter is expressed not only in its movement, but also in its generational ability to change, which is absent in inert matter (in the adaptability of organisms). Throughout geological time, living matter adapts itself slowly. It frequently and successfully overcomes obstacles to its proliferation and undergoes corresponding changes during this process.

There is another very definite and important distinction between masses of gas and living matter. The quantity of molecules for a given mass of gas is unchangeable, while the quantity of organisms for living matter is not. The "pressure" of living matter can remain unchanged in a given space, in spite of the increase of *external* pressure (external obstacles). This "pressure" of organisms is a function of reproduction, or multiplication, a characteristic feature of living matter, which is, as J. Haldane has put it, "not a problem, but an axiom." The mass of living matter in a given volume can grow to a certain limit, and its "pressure" grows correspondingly. Here we have a kind of analogy to creation, the birth of gaseous bodies through the radioactive decomposition of elements. Unfortunately, this important field of knowledge remains in the initial stages of scientific study.

Nevertheless, generations of scientists have left traces of their thought here. Studying the history of the development of scientific ideas, we find the expression of deep and diverse notions concerning the proliferation of life everywhere – its taking of the entire planet's surface. Usually these thoughts are not expressed in an adequate scientific form, but are at the stage of vague and intimate feelings. No doubt, they constitute the foundation of understanding, the "sense of nature" of several generations of natural scientists. Only gradually has Man understood the global proliferation of life on the Earth's surface.

Only at the end of the eighteenth century, great studies of the Earth's surface that had been renewed with abundant energy during the Renaissance led to our present-day concept of the world, to the understanding that life is present everywhere on Earth. The ubiquity of life is an empirical generalization that needs no other proof. We cannot explain it because any explanation inevitably leads to the affirmation that it is the manifestation of the main properties of a living organism – its multiplication. The ubiquity can be reduced to the organism's energetics, but this does not change its essence.[39] The understanding of this property of living organisms is in fact a real, though unacknowledged foundation of all our biological concepts.

At the end of the seventeenth century, it became possible to observe life that was previously invisible to the human eye – the life of microscopic beings. This discovery of A. van Leeuwenhoek (1632–1723) made a great impression on the science of that time and gave a new impetus to the idea of the ubiquity of life. But it was only in the nineteenth century that we could get a deeper understanding of its ubiquity. The discovery of life in the depths of the sea gave a notion of its quantity. The study of microbes surprisingly deepened the old notions of microscopists and the still older intuitions of empiricists such as physicians and geologists.

---

39. See V. Vernadsky, *The Biosphere*, p. 30, L., 1926

I don't think I can give a better idea of the proliferation and ubiquity of life than in the words of three great natural scientists of the nineteenth century that were uttered long before the discovery of the microbial realm and ocean life. Almost simultaneously, and independently, they profoundly expressed the idea of the ubiquity of life, this most fruitful idea of living nature, the principal reality of the cosmos. These three great natural scientists presented their thoughts far from laboratories and institutes, in the bosom of nature, far from people.

The first of them is the great German scientist, Alexander von Humboldt. After the discovery of the ubiquity of microscopic (not microbial) life by Ehrenberg, he wrote in his new edition (1826) of his *Pictures of Nature*: "When a man questions nature with the acute curiosity characteristic of him, or measures in his imagination the vast spaces of organic creation, the most powerful and profound excitement he feels is the feeling aroused in him by the fullness of life present everywhere. . . . Everywhere an observer casts his eye, he sees life, or an embryo ready to accept it."

Another scientist is our great natural scientist and thinker, Academician K. M. von Baer, a man of both the laboratory and the microscope. When he first saw the virgin nature of the polar New Land in 1838, he wrote: "No matter what point of the Earth's surface we first inhabit and no matter where we look, we see the spreading of organic life wherever Man can penetrate. And although we have not managed to reach the poles yet, the observations made at the high latitudes we could reach allow us to assert with all confidence that at the poles we shall find no limits to life, separating it from the realm of eternal death."

Almost at the same time, the third scientist – Charles Darwin – wrote in his diary on the shore of the salt lake near El-Carmen in South America: "Yes, one can, without a doubt, say that all locations of the world are supportive of life. Fresh-water lakes, underground lakes that are hidden in the depths of volcanic mountains, the sources of hot mineral water, the depths of the ocean, the upper regions of the atmosphere, the surface of perpetual snow itself – we find organized matter everywhere."[40] We express this ubiquity of life by the name given by E. Suess to the upper envelope of the Earth: 'the biosphere,' the sphere of life. The prevalence of life is a consequence of two phenomena, which are not directly connected:

1. The movement of separate "indivisible" elements of living matter, organisms;
2. The deepest mystery of life: proliferation of organisms, increase and renewal of living matter and mass.

Without going deeply into this issue now, I want to point out some geochemical consequences of this proliferation and movement performed by organisms in the biosphere.

Due to the proliferation of organisms, the whole Earth's surface is covered by a thin, mobile layer of living matter. Chemically, this living layer consists of a regular mixture of an infinite number of bodies made up of autonomous invisibles, between which an energetical chemical exchange takes place. In this living matter, two gases prevail (oxygen and carbonic acid) as well as water, proteins, carbohydrates, and fats. In many cases the layer of life includes a large number of opals, carbonates, and calcium phosphates. Often,

---

40. We have freely translated a French version of Darwin's diary which was utilized by Vernadsky. – *Ed.*

but not always, organic compounds prevail, and the quantity of water oscillates between 60% and 90%. Living matter comprises an unusually thin layer; it fills all water reservoirs and lives in all of the oceans. Oceans contain the richest accumulations of living matter; its distribution there is very uneven, but it is everywhere. The layer of living matter can practically be regarded as a continuous envelope covering the Earth.

The fact that this matter is dispersed to an extent that exceeds all the terrestrial mechanical and chemical dispersions is very characteristic. I have already pointed out the importance of this dispersion for the chemical activity of the smallest particles of matter that remain after the death of microscopic organisms. But this extreme dispersion is nothing but a consequence of living matter's tendency towards ubiquity. Considering the proliferation of living matter, it becomes clear that the increase of its mass is proportional to its surface.

This increase in mass is partly a result of gas exchange, liquid exchange, and more rarely of solid matter exchange. This exchange is always performed through the organism's surface. The increase of the organism's surface corresponds to a more intense exchange of its matter with the environment, to the acceleration of the "more or less rapid whirlpools" of life. The whole structure of living matter is determined by the power of its proliferation, which also explains the minute size of microbes and ultramarines, and the complex cellular structure of organisms.

The surface area of a certain mass of small bodies of any matter is always much greater than the surface area of the same mass in one large body of a similar form. The division of living matter into smaller and smaller organisms can to some extent be regarded as a mechanism that increases the area of living matter, intensifies its exchange, and contributes to its distribution all over the Earth's surface. It is the smallest organisms that spread with the greatest speed. Reproduction of organisms is inevitably related to the formation of complex compounds building up the organism's body. The synthesis of these compounds occurs with a speed unknown to our laboratories and factories.

Rapid proliferation of organisms, especially of microscopic ones, had drawn the attention of scientists as early as in the eighteenth and even during the seventeenth century. C. Linnaeus was already interested in these issues. He and other scientists found that one individual of *Musca carnaria* [a fly], together with its offspring, could destroy the body of a horse or a lion in a few days, leaving nothing but bones. These calculations, and Linnaeus's ideas related to them, did not receive enough attention up till the present time.

Proliferation (reproduction) of living beings, especially microscopic ones, which interests biologists from the physiological point of view, is in fact a geochemical phenomenon, as it exerts a great influence upon the quantity of living matter existing in the biosphere and controls the whole cycle of life of chemical elements. It has a still greater importance in the Earth's crust, as living matter is a source of "active" energy. Eventually, it causes movements of large masses of matter, similar to other processes in nature, such as those involving sediments of water reservoirs, accumulations created by rivers or wind, and glacial covers. This is the main geochemical manifestation of the biosphere.

The growth of geochemistry has been hindered by the absence of scientifically stated biological facts concerning this phenomenon.

Usually these facts are collected incidentally in biology, as simple examples of the speed or intensity of reproduction. In many cases they are used only to illustrate the importance of organism proliferation; the latter is spoken about as a great natural phenomenon that is usually not embraced by our minds. In geochemistry, we cannot restrict ourselves to such an approach. The role of reproduction in the migration of atoms and in geochemical processes is most significant. Having no possibility to expand on this phenomenon, I will give a few particular examples showing the power of proliferation within the confines of the Earth.

At the beginning of the nineteenth century, C. Ehrenberg gave an illustration of the amazing power of proliferation of diatoms, the silica organisms that are annually part of the sea plankton, and whose great importance in the history of silicon has already been pointed out. Ehrenberg proved that one diatom, dividing without obstacles, can give a mass of matter equal to the volume of our planet in eight days, and can double this mass during the following year.[41] An ordinary small infusorian (*Paramaecium*) can yield a mass of protoplasm in five years with a volume that is 104 times larger than the Earth's volume.

The realm of microbes, unknown to C. Ehrenberg, gives still more amazing examples. A single bacterium in a favorable nutritive medium, can in one day give over 10^{25} indivisibles (i.e., septillions) of individuals. A microscopic trace remains of its initial matter, around which masses of the same chemical matter concentrate, consisting of the most complex organic combinations. According

---

41. I won't try to duplicate C. Ehrenberg's calculation, except for the following: If the diatom and each of its offspring divided once every 90 minutes for 8 days, it would produce over 10^{38} diatoms. If that equals the volume of the Earth (and I have no idea if it does), then it would double the Earth's volume in the next 90 minutes – it certainly would not take a year. –*Ed.*

to F. Cohn, a bacterium gives an offspring of 10^{36} individuals during four and a half days. This amount could fill the ocean, and its weight would obviously exceed that of the latter, as the specific weight of a bacterium in the ocean, as far as it is known, is a little greater than the weight of salt water. This means that in three days, a living particle of a most complex chemical composition and weighing several trillionth of a gram (10^{-11} to 10^{-10}) gives a mass of growing matter of the same complex composition weighing more than 1.4×10^{17} t; that is, several thousands of times more than all the organisms existing in the biosphere at any given moment.

This is not the extreme limit of creation of matter by living organisms' geochemical energy, or of the amazing whirlpool of migrations of elements they can bring about. The amplitude of possible oscillations is very wide. If cholera bacteria can cover the whole planet's surface with a continuous layer in a day and a quarter, the terrestrial organism with the slowest reproduction rate – the Indian elephant – will do so in 3,000 to 3,500 years. The speed of transfer of the geochemical energy of life is about 33,000 cm/s for a cholera bacterium, which is close to the speed of sound waves in the air. For an elephant this speed will be about 102 cm/s.

In the measureless geological duration of terrestrial life, the geochemical effect of both morphological forms – an elephant and a bacterium – will finally turn out to be the same. The most slowly reproducing organism in the biosphere, as well as the quickest one, will be a great force changing the environment. These figures give an idea of the potential for changing geological processes inherent in organisms. In reality it manifests itself in the environment to a smaller extent, without reaching the limit, but the order of the phenomena remains almost the same.

But sometimes we see its manifestations in full swing. We observe these phenomena every day, being unaware of their magnitude. According to the beautiful old image that has recently been returned to science by W. Hudson, they are "waves of life" – sudden explosions of life. Sometimes, during several days or several hours, we see how myriads of living beings appear: algae, insects, spiders, small verte-brates, and other animals. In a very short time, and under favorable conditions, living matter prepares a great quantity of proteins, fats, and carbohydrates.

Thus, in the Adriatic Sea, a phenomenon is sometimes observed by the Italians called "mare sporco": a sudden, vast multiplication of diatoms resulting in accumulations of diatoms poor in silica. These diatoms multiply with such rapidity that the amount of silicon dissolved in the seawater has no time to be restored. Their assemblages appear and cover the sea surface with a layer of several meters. After several days of such proliferation, they sink to the bottom as jelly-like masses, which quickly decompose or get eaten. The broken balance of nature is soon restored. Similar phenomena can be observed on a still larger scale related to ocean plankton and its coastal "life conglomerates." Here, plankton cover areas that are impossible in sea or on land. These phenomena on our planet take place continuously; if they die away in some place, they go on in another.

The same phenomena can often be observed on land too, such as the "blooming" of lakes, rivers, marshes, and pools. Plants like duckweed (*Lemna*), algae, unicellular animals, and other organisms, suddenly cover the water surface with a more or less continuous layer. They renew periodically, quickly creating vast quantities of organic compounds. But the balance is soon restored again, as they decompose and tracelessly enter the cycle of life. If the decomposition is not complete, they can, as we have seen, give birth to coal and crude oil. The phenomenon itself fully corresponds to the great potential forces whose numerical expression I gave above.

The role of these phenomena in the economy of nature is enormous. These explosions of life have a definite existence – they are short. Soon the balance is restored and nature regains its usual image. These phenomena manifest themselves every moment in many areas of our planet. They are indeed "waves of life," which are always existent and which point out the mechanical feature of the phenomena. Their existence must be connected with the very structure of living matter. Nevertheless, it attracts little attention from the scientists of our century. Scientific mind turned to it more than once when ideas of harmony and economy of nature were dominant. I think our great natural scientist K. M. von Baer approached this phenomenon 60 years ago as "a great echo of the planet."

10 the quantity of carbon in living matter

The quantitative expression of the geochemical effect of proliferation may be given by the formation of living matter – its mass in a certain area of the biosphere. This quantity is determined by green plants, the only beings capable of capturing the radiant energy of the Sun. The quantity of plant mass on a definite area is restricted. A notion about the order of its size can be arrived at by studying the harvests of our agriculture. For instance, the maximum corn harvest in England can give more than 30 tons of organic matter per hectare and per reaping. But this is an exceptional example. In general, "taking the best conditions and the best cultures," the harvest does not exceed 15 tons per hectare according to Duclaux.

In tropical countries this quantity is of course much larger. At banana plantations, the achieved harvest has long been about 50 tons of dry organic matter per hectare, without considering the weight of branches and leaves of these Musaceae. But even here we are apparently too far from achieving maximum fertility of the soil. Manioc – *Manihot utilisissima* Pohe of the Euphorbiaceae family – gives still larger quantities of produce; about 250 tons of fresh tubers per hectare, without considering the above-ground parts of the plants. Moss and lichen conglomerates of wet tundras of the arctic and subarctic countries, as well as wet northern meadows, yield similar quantities of living matter; maybe larger than the harvests of our countries.

The forests of our latitudes annually give an increase in timber and leaves up to 7.5 tons per hectare. In tropical forests, the mass of living matter per hectare must be larger still. The numbers correspond to carbon masses ranging from 6–7 tons to 50–60 tons per hectare. High as these numbers may seem to us, they surely give only minimal parameters for a full calculation. They never correspond to all the carbon gathered by living matter per hectare; they only show the part useful for humans. Man is never interested in all living matter of the soil. In cereals, roots are never taken into account; if they were, the total organic mass should be doubled. The realm of microbes and animals of soil and subsoil is never regarded. The quantity of life Man ignores is apparently as large as the quantity of organic matter he uses for his needs. At any rate, it is of the same and maybe of a much larger order.

The harvests gathered from the fields of our climate do not correspond to the organic masses forming in tropical and subtropical countries, or those in our peat swamps, marshy tropical forests, and in wet prairies of subarctic countries. Besides, a considerable part of carbon is concentrated in the animal realm, especially in that of the Crustaceae. It has long been obvious that living matter, containing at least 60% of easily released water, concentrates in areas on Earth in proportion to the quantity of water present.

The maximum bloom of life manifests itself in the waters of the hydrosphere. There are two sources for the formation of the initial mass of living matter: firstly, the plankton of the sea surface, which is mostly microscopic zooplankton and phytoplankton, and secondly, the algae, zoster, and other water plants in the benthos, coastal and sargasso conglomerates of life. All sea life is based on a substratum of green plants. In the ocean, animal organisms prevail that are rich in nitrogen. They dominate not only in the diversity of their forms, but apparently also in their quantity and weight. All carbon contained in sea life has ultimately been derived from green living matter, and almost exclusively from plants.[42] Due to their intense metabolism, rapid proliferation, and mutual consumption, sea animals exhaust one and the same organic store of carbon extracted from the air by plants.

In the hydrosphere, life exists in a liquid layer with an average thickness of 3.8 km, while on land this layer is no more than a few meters. This results in the fact (while the ocean covers an area 2.4 times larger than the continents) that the major mass of animal and plant carbon is concentrated in the ocean, most probably distributed unevenly. Another phenomenon observed on land – the concentration of living organisms in definite layers (in the soil and nearby) – is still more pronounced in the ocean and in other aquatic reservoirs of life.

---

42. Green plants make up half of all living matter. See: V. Vernadsky, *The Biosphere*, p. 95. L., 1926.

Together with areas of high life concentration, in which geochemical processes are intense, we can distinguish areas so poor in life that its influence dies away. In the oceanic waters we can distinguish two life films: the plankton film on and near the surface (50 to 60 meters thick) and the bottom film, which is usually much thicker. Besides, we must distinguish two other kinds of "life conglomerates" in the ocean: the sargasso conglomerates on the surface of deep waters and the coastal concentrations on the border of sea and land, such as near coasts and over shoals. Such a distribution of life presents a picture of the hydrosphere (and inland water reservoirs) that has existed in all geological periods. The way the geochemical energy of life and the migrations of elements it brings about are distributed in the Earth's crust are perennial forms on a geological timescale.

The quantitative characterization of plankton developed by V. Hensen more than 40 years ago finds more and more organized beings with the improvement of research methods. The newer definitions by E. Allen point out that more than 464,600 minute organisms, excluding bacteria and nanoplankton, exist in a liter of a plankton film in the Atlantic Ocean. For the Atlantic Ocean of our latitudes, the characterization of phytoplankton alone, excluding the microplankton and animal plankton that exceed it in its richness, gives annually 10–15 tons of dry organic matter per hectare. According to Petersen, organic matter in the form of floating plants – such as zoster – and their remains, corresponds by weight to $4.1 \times 10^{-3} - 1 \times 10^{-2}$ % of the mass of seawater in those concentrations that are richest in life.

To compare this with soil productivity, we can express the total quantity of living matter of the hydrosphere by calculating it per hectare. It turns out that the quantity of visible life in the hydrosphere is a little larger than its quantity in the terrestrial areas where life is abundant. The maximum harvests on land and the maximum harvests in the sea are comparable in their amount of dry organic matter per hectare, although the fertile layers are distinctly different in thickness. The thickness corresponding to a hectare of sea reaches an average of several kilometers, and that of a corresponding layer of terrestrial life, in its maximum expression – for instance in the forests – can sometimes reach a hundred or more meters. Usually it reaches several dozens of meters or less.

The fact that the harvests of terrestrial and marine green plants and animals show similar figures, when calculated per hectare, is apparently explained by living matter being created by the Sun's radiant energy; that is, depending upon the surface of the illuminated area, which is the same both for the land and the sea. This issue is of great theoretical interest and is most closely connected with the geochemical regime of the planet. The contemporary state of science allows us to touch upon it only casually. Following are some of its manifestations.

P. Boyesen-Jensen described the annual yield of living matter for sea bottom animals of the benthos near Denmark. The harvest changes from year to year between 42.1 and 77.1 grams of carbon per 1 m². This corresponds approximately to 206.3–377.8 kg of carbon per hectare. Besides, Boyesen-Jensen regarded only benthos and ignored plankton, microbes, and mobile animals. The benthos he studied lives at the expense of zoster and its remains. Zoster gives 12 tons of dry organic matter a year per hectare. The animal world lives on its remains and its living matter, which has no time to reach maturity. A balance is created: animals, mainly nectonic fish, consume the zoster that quickly covers the losses by new growth.

Unfortunately we cannot give exact figures for the entire quantity of living matter in the biosphere and the carbon it contains. All we can do is show the approximate order of the figures corresponding to it. But such an imperfect approximation is better than complete ignorance. It is evident that the quantity of carbon related to life is limited, and that its total quantity in comparison to the mass of carbon contained in the Earth's crust is rather small. As we have seen, this mass is close to 1×10^{17} t, while all the carbon of the Earth's crust up to 15–20 km makes up 0.5%. This is not the maximum figure, and I think that it must increase with a more exact calculation of gases, dispersed organic matter, and carbon of massive rocks. Possibly it will surpass 0.5–0.6%.

The mass of all living matter is hardly much more than 0.1% of the Earth's surface, i.e., 2×10^{16} t, and most probably it is smaller, since by assuming 100 t of organic matter to correspond to the amount of living organisms per hectare, we arrive at 5.1×10^{12} t for the whole biosphere. The order of this figure can hardly be changed by living matter, taking into account all the animals and microbes of soils and muds. It probably corresponds to a value of $n \times 10^{13}$, while a figure of the order of 10^{14} will correspond to its limit. The average quantity of carbon in living matter cannot exceed 7%, and actually must be much smaller, taking into account aquatic organisms. This results in the fact that the quantity of carbon captured by living matter cannot exceed $n \times 10^{13}$- 10^{14} t and is most probably less than that. Thus, it corresponds to several millesimals [thousandths] of the Earth's total carbon mass.

But its actual significance is incomparably larger. Living matter presents a dynamic balance of atoms, which mass is in intense migrational movement. Part of the carbon atoms is released from the organism and immediately replaced by other atoms. Millesimals give no notion about the significance of living matter's carbon; the atoms' *nature* is much more important. During a short time, for instance in a year, living matter creates a movement of carbon masses probably on the same order as the entire amount of terrestrial carbon.

The facts we have noted in the history of silicon and oxygen can be repeated here too. This does not mean of course, that all the atoms of carbon actually pass through living matter. A part of the atoms is continuously located in it; atoms return to living matter immediately if for some reason they leave it. Only a small part of carbon leaves living matter for long periods of time, and sometimes for millions of years (in the case of crude oil, carbon, limestone etc.). Some scientists, using other ways of calculating, arrive at smaller figures for the quantity of living matter existing on our planet at any moment. But the difference is not very large. The most probable of the smallest numbers indicates $n \times 10^{11}$ and maybe $n \times 10^{12}$ tons for terrestrial living matter.

11   the constancy of the mass of living matter

The exact characterization of living matter and of its carbon is of great importance also for other reasons besides its influence on geochemical processes. It is obvious that any crucial change of mass or composition of this living matter must be clearly expressed in the history of chemical elements, the composition of inert matter, and the environment in which an organism lives. But does such a change occur, and can it occur?

Studying the history of the Earth's development, we come across a fact of great importance, the consequences of which have usually been ignored: the constant chemical nature of the Earth's crust throughout the whole course of geological time. No doubt, the minerals formed over geological time have always been similar. Everywhere and always, not only since the Cambrian era, but even since the Archean eras, the same minerals form; there is no change whatever. Not only the minerals themselves remain unchangeable, but also their paragenesis, as well as their mutual quantities.

It must therefore be concluded that geochemical phenomena have taken place without considerable changes since the Archean eras. This also results in the fact that the average quantities and composition of living matter have remained approximately the same during this time, unfathomable in its duration. Otherwise, due to the importance of organisms in the geochemical history of all chemical elements, neither minerals nor their compounds could remain the same all the time. Thus, the quantity of living matter is apparently a planetary constant since the Archean era (i.e., during the whole course of geological time).

Regarding the complex dynamic balances we see in the biosphere, speaking about the *constancy* of the phenomenon, it is most likely impossible to think that the given phenomenon does not change in its numerical value. It can only be asserted that the *limits of the oscillations* do not change. We are used to such a form of constancy in the compositions of atmosphere and ocean water (salts). Their constancy, that is, the unchanged limits of oscillations, shows that the biosphere exists in a very stable equilibrium. We are used to the stability of the mentioned compositions within the limits of historic time, but this stability must show us that we are dealing with a phenomenon that cannot change without destroying some very deep characteristics of nature.

At present a very probable hypothesis can be formulated, which states that all these constant limits of oscillations are geologically eternal (i.e., unchangeable over geological time) and constitute the principal features of the biosphere's structure. The facts known to me now contain no contradictions to this hypothesis. It is interesting to note that here we are returning to ideas almost forgotten in biology, but with a very interesting history. These old biological ideas will have to be revived, for it is obvious that the mentioned constancy of the mass of living matter is important not only from the standpoint of geochemistry, but also from that of biology.

Buffon was the first to pose a clear question in biology about the quantity of "life," as they put it then, existing on our planet. He presumed this quantity to be unchangeable, and living matter to originate from particles – organic molecules different from the particles of inert matter. These organic "molecules" were in his opinion immortal and unchangeable. After an organism's death they enter other organisms, and thus they are moving from the beginning of eternity. These bold ideas resemble very old notions of Indian thought, but it can hardly be supposed that ideas from India influenced Buffon directly, as his work had begun earlier than their arrival in Western Europe. The genesis of these ideas is complex and not quite clear. They must be very ancient and revived in a new form. We find their traces in the philosophic and scientific atmosphere of the Renaissance, and they can be clearly traced up to the seventeenth century. Their actual roots, however, must be looked for in the ancient religious and philosophical notions of Asia.

From Buffon's ideas it logically follows that the quantity of organic matter – the matter of all the organisms on Earth in the biosphere that corresponds to the mass of the eternal organic molecules – is constant. These ideas of Buffon have never been forgotten. However, they represented a trend of scientific thought that has remained outside the mainstream trend. We observe their revival in new forms in the studies of such naturalists of the nineteenth century as L. Ocken, J. Sniadecky, P. Flourens, K. M. von Baer, L. Agassiz, W. Preyer, E. Von Gartmann, A. Brandt, and others. At the end of the century, the German physiologist W. Preyer even tried to deepen this idea and to introduce it into science, but he failed. In general it was supposed that the quantity of living matter (i.e., of life, something like the store of life) remains unchanged throughout geological time. K. M. von Baer, however, allowed the possibility of its increase over the course of time.

These half-forgotten theories are arousing new interest at present. The quantitative study of life phenomena has acquired such great significance now that its influence causes considerable changes in the ideas created in a time when qualitative and morphological work reigned supreme, which ideas aroused no doubts up till recent times. These old ideas have found a favorable soil today; their time has come. The constant quantity of living matter on Earth – the constant range of its oscillations – is an empirical fact and a necessary consequence of the totality of geochemical facts.

12   chemical structure of the earth's crust

These phenomena have one more consequence. Not only the quantity, but also the average chemical composition of living matter must be constant in historical as well as geological time. From the geochemical standpoint, this constancy arouses still less doubt than that of the *quantity* of living matter, for every chemical change of living matter should have manifested itself in the formation of new minerals or in the increase and decrease of its quantity.

Meanwhile, throughout geological history, minerals always remain the same. Periods of intense coal formation, for example, have repeatedly occurred on a geological timescale. The same is true of minerals having no relation to life. For instance, the appearance of several glacial periods throughout geological history corresponded to an increase of the quantity of ice. We are also aware of several repeated penetrations of native gold or cassiterite to the biosphere and the stratisphere. These periodic changes in the quantity of certain minerals obviously depend on some other reasons than those determining changes in life phenomena.

We must see them as periodic oscillations of physical and geographical interactions, connected with the phenomena taking place in the *depths of the Earth's crust rather than the whole planet,* or connected with oscillations in the movement of magmatic centers, which are related to orogenic phenomena and radioactive processes. Such periodic oscillations of both the chemical composition of living matter and of its quantity are very probable, as the general reasons mentioned above must affect them too. But we do not consider them now, since we do not yet realize the constancy of the average composition and quantity of living matter that must precede them in our scientific understanding.

In this respect it is necessary first of all to note and to consider the unchangeable nature of all forms of terrestrial inert matter, of all most diverse natural chemical compounds, and of all minerals that are related to living matter. This is a geologically everlasting phenomenon. Their unchangeable character is based on the chemical composition of living matter being constant and definite, and remaining, on average, unchanged throughout geological time. In this field, the conclusions of geochemistry seem to contradict the data of biological sciences, namely of paleontology.

The most distinct empirical generalization dominating all our ideas about the Earth's realm is the incessant and regular change of the *morphological* structure of living matter – the evolution of species. The idea of the evolution of the organic realm seems one of the greatest achievements of the human mind in the last century. It is obvious that morphological change must be closely connected with chemical change, as the form of an organism, like the form of any physical body, is determined first of all by its internal chemical structure.[43]

These two facts – the morphological evolution of life throughout geological time, every change of which is connected with a chemical change, and the unchangeable character of its average chemical composition within defined limits, are firmly stated empirical facts. Their seeming contradiction requires explanation. As a probable hypothesis, it can be stated that the chemical change necessarily accompanying every morphological change, formation of a species,

---

43. Vernadsky would be interested to learn that the internal "chemical changes" that influence the morphology of an organism in a primary way are as subtle as the *sequences* of nucleotides in nucleic acids (DNA & RNA), which in turn control the sequences of amino acids in protein enzymes. The enzymes in turn affect chemical reactions that finally result, in ways not yet understood, in the observed changes in morphology. *–Ed.*

race, etc., takes place within chemically unchangeable limits of diverse living matter.

Living matter's total composition has always remained the same throughout geological time. Separate chemical changes are compensated by simultaneous changes of the opposite direction in other homogeneous matter. If this explanation is correct, it must be acknowledged that the structure of living matter as a whole has characteristic unchanging features in the course of species evolution. Here I cannot plunge into the analysis of those important issues, which are of great scientific interest. I shall only observe that the conclusions, following from the unchangeable nature of the average composition of living matter throughout geological time, can and must be seemingly contradictory to the morphological (and chemical) evolution of organized beings.

The chemical composition of living matter is known to us only in its most general features and to a very insufficient extent. Up to recent times, there was no complete analysis of any living organism, which in its precision and correctness, could be compared to the contemporary chemical analyses of minerals or rocks. At present, a small number of such analyses have been carried out and partly published by the Biogeochemical Laboratory of our Academy of Sciences. The work is systematic but slow. Only when it leaves the limits of this small institution, and circulates widely amongst biologists who realize its importance, will the situation change. Now, a geochemist must work in two fields that are equally necessary for him (in the field of inert matter and in that of living matter) with scientific facts of different precision. In biogeochemical issues, he has to draw scientific consequences from a large quantity of incomplete and insufficient facts.

There are thousands of incomplete and non-equivalent analyses of parts of organisms, or their total mass, dried at different temperatures up to 125 °C. The main mass of the organism, the water it contains, is known to us less exactly than its other components. The volatile parts, as well as the gaseous matter forming in the process of drying below 100 °C, are always lost and not taken into account – they are not reflected in the obtained figures. There are hundreds, or maybe thousands, of analyses of animal and plant ashes in which the correlation of the ashes to the average mass of the living organism is not mentioned. On the whole, the average mass of a living organism is studied very poorly, although this figure is absolutely necessary for all geochemical conclusions. The gaseous part of the organisms has not been taken into account, although there is no doubt about its foremost importance and the fact that it changes the average figures of living matter's composition.

The overwhelming majority of analyses have been made for particular purposes of interest to biologists, or for practical issues of interest to zootechnicians, physicians, or agriculturists. We must look for separate geochemical data in this chaotic mass of facts collected over more than 100 years as the result of a vast amount of scientific work. No doubt, in recent years the situation has changed. The amount of exact quantitative data for separate elements is increasing rapidly. Apparently, in a few years we shall have sufficient data for the approximate quantitative definition of the chemical composition of living matter. Within the next few years it will be possible to analyse the main classes of organisms for the majority of chemical elements. But at present no exact quantitative notion of the elements composing living matter can be arrived at.

Before we can achieve a level of precision comparable to that of the chemical composition of minerals and rocks – the lithosphere – quantitative information must be obtained for the largest possible number of separate species, families, and biocoenoses. Still, at present it is possible to give the order of the expansion of some chemical elements in living matter. Some more extensive empirical generalizations can be expressed, as was recently done by A. Vinogradov (1933)[44], who has shown the dependence of chemical composition on the Periodic System of Elements. The obtained curve is distinctly different from the similar curve for Clarke's number – for the composition of the lithosphere.

In table 9, I give the results of such a definition by placing the elements in a diminishing order regarding mass.

table 9   average content of chemical elements in living matter

Groups	Mass in Tons	Elements
1	$>10^1$	O, H
2	10^0-10^1	C, N, Ca
3	$10^{-1}-10^0$	S, P, Si, K
4	$10^{-2}-10^{-1}$	Mg, Fe, Na, Cl, A, Zn
5	$10^{-3}-10^{-2}$	Cu, Br, I, Mn, B
6	$10^{-4}-10^{-3}$	As, F, Pb, Ti, V, Cr, Ni, Sr, Li
7	$10^{-5}-10^{-4}$	Ag, Co, Ba, Rb, Sn, Mo
8	$10^{-6}-10^{-5}$	Au
9	$10^{-7}-10^{-6}$	Hg
. . .		
13	$10^{-12}-10^{-11}$	Ra

---

44. A. Vinogradov, Act. Labor. Biogechem., 3, L., 1934

I first presented a table in 1922.[45] The comparison of table 9 with that one demonstrates the growth of our knowledge in 12 years. Each row of the table contains chemical elements with quantities of similar percentage. Calculations can also be made for separate large classes of families and orders of organisms, though for rather few of them. For instance, they can be given for insects, higher chlorophyll plants, algae, etc., but they cannot be given for fish. But even in this general form it could help to quantitatively take into account the influence of living matter upon its environment, if we could calculate the composition of life's environment – the biosphere – with the same, limited accuracy. The average chemical composition of the biosphere is as yet unknown. The only thing we know exactly is the composition of the lithosphere, which is quite different from that of the biosphere. In comparison, the latter must be enriched in carbon, nitrogen, calcium, potassium etc., as it is completely embraced by living matter and organogenic minerals.

Regarding sedimentary rocks – to say nothing about silts and soils – the percentage of carbon sometimes reaches 1–2% in non-carbonate rocks, and in carbonate rocks its content exceeds 12%. The

---

45. The units for tables 9, 10, & 11 don't make much sense as they stand. table 9 presents various numbers as the mass in tons. By itself, this is meaningless; it must be tons per some volume of "living matter." (There is no heading for the table, but the text suggests that we are dealing with "living matter.") Yet the text suggests that the figures of table 9 are percentages – percentages of some mass of living matter? The figures of table 10 are said to represent mass percent (of sea water), but the total would not exceed 25%. Probably Group 1 should be $> 10^1$ (which is >10). In the figures for table 11, the term concentration must mean that the element is concentrated by the factor expressed by the figure. For example, K is concentrated $n \times 10^1$ times by the organism as it is absorbed. Even if the units in these tables are not clear, Vernadsky's goal is to compare the various elements with each other, which is easily done in the tables. *–Ed.*

composition of the lithosphere on average corresponds to that of acid igneous rocks and not to the composition of the environment in which life processes occur. Although we cannot state the average composition of the *biosphere*, it is obviously much *closer* to the average composition of living matter than that of the lithosphere; this change has been created by life.

table 10 average composition of the hydrosphere

Groups	Mass %	Elements
1	10^1	O, H
2	10^0	Cl, Na
3	10^{-1}	Mg
4	10^{-2}	S, Ca, K
5	10^{-3}	C, Br, N
6	10^{-4}	Si, Rb, Fe
7	10^{-5}	P, F, I, B, Ar, Cu, Ag, U
8	10^{-6}	Li, Th, As
9	10^{-7}	Zn
12	10^{-10}	Au
14	10^{-12}	Ra

What cannot be done for all organisms can be done for a majority of them. For marine organisms, this comparison can be made with a certain exactitude. The average composition of the hydrosphere according to the present state of our knowledge can be presented in groups as has been done in table 10. The comparison of tables 9 and 10 gives us valuable information about the chemical work of living matter. Ocean water is a chemically very definite body: a strong solution, mainly unchanged throughout geological time. But this absence of change must concern only the average composition of its salt.

In the areas where life is concentrated, the composition of seawater changes more or less considerably under the influence of the geochemical energy of life.[46] As far as we know, life is unevenly distributed in seas and oceans; it is sparse in some parts and dense in others. In its vast concentrations (constant or temporary, and in certain areas of the ocean), living matter can make up several percent of the total mass of water. Since its composition is very different from that of seawater, these parts of the hydrosphere taken as a whole are distinctly different from the rest.

This statement makes it necessary to revise the notions about the functional dependence of life phenomena on their environment – the biosphere – and to consider them in their biogeochemical aspect. Among such notions that mistakenly simplify the chemical relations between living and inert media, there are widespread notions about the similar composition of mammals' internal liquids and that of seawater. The analogy between the compositions of these two liquids, quite far-fetched and incomplete, was a consequence of the fact that the ancestors of mammals were marine organisms.

In particular, attempts were made to explain the composition of blood. In the salty part of the solutions' composition, an analogy was found with the composition of seawater. From the biogeochemical standpoint this is most improbable, as a terrestrial organism depends for its composition on the terrestrial environment, and it cannot retain the composition of the sea, which is alien to it. Biologists have now come to the conclusion also that blood composition always changes in organisms, when passing from the sea to the dry land (W. Daukin).

---

46. V. Vernadsky. *Living Matter and the Chemistry of the Sea*, p. 8 and further, P., 1923.

Direct observation also shows that seawater's composition is insufficient to explain that of an organism's liquids. Without taking into account the biogeochemical character of the phenomenon and the organic connection between a terrestrial organism's composition and the environment, outstanding scientists of the nineteenth century allowed the hypothesis that the ancient ocean had a different composition. But this new hypothesis contradicts the geological and geochemical facts. There is some similarity in the compositions of both liquids, but the difference between them requires another explanation, which I cannot dwell upon here.

Living matter in its totality (table 9) has quite a different composition than that of the salty oceanic mass. Living matter dispersed in seawater is in a state of constant, very active chemical exchange with it; it is constantly giving seawater its atoms and extracting new ones from it. We know that this is how the regular enrichment of living organisms with separate elements takes place. This fact follows already from tables 9 and 10; living matter extracts certain atoms from the salty water of the ocean.

But before making this comparison, it should be taken into account that living matter borrows only a part of its composition from seawater. It extracts nitrogen and carbon directly from atmospheric gases, or from gases dissolved in the water. It is an apparatus increasing the content of nitrogen and carbon in the sea, by introducing its atoms there. The comparison between tables 9 and 10 shows that living matter enriches the ocean with the following elements as presented in table 11.

table 11   concentration of individual elements by living matter as compared to ocean water

Elements	Concentration
F, B, K, S	$n \times 10^1$
Fe, Br, Sr, As, Ag	$n \times 10^2$
Si, P	$n \times 10^3$
Cu, Ca, I	$n \times 10^4$
Zn, Mn	$n \times 10^5$
Living matter concentrates Na, Cl, and Mg as well, but not on the order of 10^1	

The heterogeneous living matter of the ocean – sea life taken as a whole – can be regarded as a special mechanism that completely changes the chemistry of the sea. It affects all ocean chemistry. It influences the precipitation of sea silts and vadose minerals that make up the sea bottom, and the sedimentary rock minerals originating from them are also related to it. Life's influence expands even further to the stratisphere and the metamorphic geosphere. The vadose minerals undergo subsequent changes, and the energetic factor bringing about this change is the geochemical energy of life. In these processes of metamorphism, the vadose minerals become phreatic. Such a biogenic process, as we have seen, clearly manifests itself in the history of petroleum.

Taking into account these post-life energy manifestations, the influence of marine living matter upon the chemistry of the ocean, and especially of specific ocean regions or of inland seas, is still much more profound than that of the processes related to terrestrial living matter. The migration of atoms (tables 10 and 11) gives a too general and insufficient idea about the biogeochemical role of the organisms of the sea, since their post-life effect is not taken into consideration.

13 organisms: concentrators of chemical elements

Separate species of plants and animals (homogeneous living matter) are often distinctly different in chemical composition both from one another and from the average composition of living matter. Local agglomerates of homogeneous living matter – certain biocoenoses that differ most from the average composition of living matter – are of great importance not only in the whole picture of geological phenomena, but especially in the geochemical history of particular chemical elements.

From this standpoint, homogeneous living matter (species, races, etc.) can be distinguished by a different concentration of chemical elements. The extreme enrichments in chemical elements are especially pronounced when a chemical element is contained in an organism in quantities exceeding 10% of the living organism's mass (corresponding to the concentration of the first group of table 9), or when its quantity varies between 1 and 10 percent of this mass (corresponding to the second group of table 9).

The state of our present knowledge can be represented by the following two rows of chemical elements:[47]

1. Chemical elements of quantities exceeding 10% of the mass of the whole living organism: In all organisms this is O and H, and in some species C, Ca, Al, Fe, Si, Mg, Ba, S, Sr, P? Mn? K?
2. Chemical elements of quantities varying between 1 and 10 percent of the organism's weight: In all organisms this is O and H, and in some species only C, N, Ca, Si, Al, Fe, Mn, Mg, K, Na, S, Cl, Zn, P, Br, I, and probably Cu and V.

47. V. Vernadsky, *Living Matter and the Chemistry of the Sea*, p., 1923.

Thus we know at least 10 to 17 elements that concentrate in living matter. This concentration is still more important due to the fact that these organisms – concentrators of specific chemical elements – are not rare bodies in nature. On the contrary, such organisms are numerous, have considerable masses of matter, and thus play a considerable part in the economy of the biosphere. We have already mentioned this fact for ferrobacteria and manganese bacteria in the geochemistry of manganese, for algae and sponges in the history of iodine, and for diatoms and radiolarians in the geochemistry of silicon. All plants and animals on land concentrate nitrogen, carbon, and often calcium.

Through special organisms that are concentrating chemical elements, living matter regulates the chemistry of these elements in the Earth's crust. This is not an incidental phenomenon; living matter of a certain composition plays the main role in the chemical mechanism of the Earth's crust. The chemistry of organized beings is so poorly studied that new discoveries can always be expected in this field; for instance the strontium, barium and phosphorus organisms, or those rich in Mg or Zn, were only recently discovered.

The mineral composition of the Earth's crust being closely connected with living matter and remaining unchanged means that different organisms concentrating the same chemical elements had to exist throughout geological time. With evolution, with the appearance of new species or genera, this geochemical function of life had to remain unchanged. The enormous resistance of the environment, the organized character of the biosphere, poses insurmountable limits for the process of evolution to "direct" it. With the extinction of a concentrating species, it has to be replaced by another, performing the same biogeochemical function; for instance a silicon organism being replaced by another silicon concentrator.

The influence of living matter is very great not only in the history of elements whose concentrating organisms we know, but in many other cases of geochemical processes. For instance, elements should be mentioned that are closely related to free oxygen, a product of life, regarding their geochemistry. The geochemistry of the majority of elements depends on living matter. This influence has definitely been proven for all the cyclic elements and for many dispersed ones: 48 elements in total. It is probably also true for certain rare elements. Thus, the whole existence of the Earth's crust – at least 90 % of its mass – depends on life regarding its geochemically significant characteristics.

14   carbonic acid as the main source of carbon in living matter [48]

A considerable portion of living matter originates from the carbonic acid of the atmosphere or the carbonic acid dissolved in water, for these are the only sources from which it extracts the necessary carbon. All other sources of carbon accessible to living matter in the Earth's crust, and even a large part of the carbonic acid itself, are genetically connected with living matter. However, volcanic

48. Carbon dioxide, CO_2, dissolves in water, becoming carbonic acid, H_2CO_3, in which form it penetrates living cells. Vernadsky's use of "carbonic acid" in this and the next section would now be replaced by "carbon dioxide" or simply CO_2 when he is talking about the CO_2 in the atmosphere. Dissolved in water, it is indeed carbonic acid. Actually, in Vernadsky's time, many authors referred to CO_2 as *carbonic-acid gas* or *carbonic anhydride*. This would be incorrect today. –Ed.

emanations, gases of mud volcanoes and many mineral sources, and gases connected with tectonic movements contain two other gases – CH_4 and CO – which can originate independently from life, although their quantity is inconsiderable. There are organisms that are capable of living at their expense and that use them for the formation of carbon compounds, but they play quite a nonessential role as compared to the main mass of carbon in living matter, which originates from carbonic acid.

The obviousness of these facts was discovered at the beginning of the last century. The importance of the atmosphere for the nutrition of a green plant and for the formation of its body has been acknowledged since the end of the eighteenth century, after Lavoisier's, Priestley's, Cavendish's, Senebier's, and Ingen-Housz's studies of gases. Apparently, it was Ingen-Housz who first understood the role of gaseous carbonic acid. Now we know that the biogenic role of the atmosphere was predicted by scientists long before the appearance of major works at the end of the eighteenth century on gases. As early as in the seventeenth century, the existence of leaves as nutrition organs was perceived. In 1723, based on these directions, S. Hales saw a gaseous component as the matter necessary for a plant's life. D. Mackbridge pointed at carbonic acid as being such a component in 1776, but neither he, nor Black, who studied the role and characteristics of carbonic acid in nature, had any ideas about its composition, about the presence of carbon in it. This fact was discovered only by Lavoisier, who created quite new notions about the history of carbon in the biosphere.

With the appearance of the great works of T. de Saussure from Geneva (1797–1804), the importance of the atmosphere's carbonic acid as a source of carbon for living matter was firmly established.

But de Saussure had doubts – he did not understand the significance of his discovery. In the carbonic acid of the air he saw only the *main* source of the organisms' carbon. The significance of the carbonic acid of the atmosphere as the only source of living matter's carbon was really understood only 20 to 30 years later. Thoughts about the dominant role of atmospheric carbon for the formation of an organism's body had penetrated science much earlier, at the beginning of the nineteenth century, in the theoretical constructions of our Academician G. Parot and the French botanist A. Brogniart, which have been proven to be wrong. However, the basis of their theories was correct.

Only after 80 to 100 years of scientific work, and after the studies of Boussingault in the period of 1830–1840, was the absolute role of carbonic acid in the formation of living matter proven. Since then, only small corrections have been made to this idea; for instance, microbes have been discovered that are capable of doing without CO_2 and using the carbon of methane and carbon oxide instead. More complex life reactions of aquatic organisms have also been discovered. In 1833, Raspail proved the ability of aquatic organisms to use the carbonic acid of bicarbonates, and also, the carbonic acid of soil solutions. The total quantity of carbonic acid in the air and in the water accessible to terrestrial plants exceeds 10^{14} t and may reach 10^{15}.

Carbonic acid, like water, is a component of the highest consequence in the Earth's crust. Water and carbonic acid, placed onto the surface of our planet by the blind forces of cosmic evolution, serve with maximum force to make stable, complex, and of long duration, both the living organism itself and the world surrounding it. This idea, expressed by L. Henderson, gives a clear notion about one of the most characteristic facts of our planet's structure. Carbonic acid, like water, is a compound with unique properties, and it is as adapted to life as life is adapted to water. There would not have been any life without carbonic acid. Carbonic acid is introduced into living matter mainly by green chlorophyll plants. In nature, there is a close and complex relationship between living matter and carbonic acid. The whole organism is adapted to extracting and processing carbonic acid.

The capacity for extracting carbonic acid from the gaseous medium, and for processing it into carbonaceous compounds, depends on many conditions and is different for every organism. It changes in close connection with the ecological requirements of an organism; it is different for Sun and shade plants, and for various *classes and groups* of plants, maybe due to the character of their chlorophyll. It also changes under the influence of light intensity, temperature, and the quantity of carbonic acid in the environment. On the whole, there can hardly be any doubt that green organisms can extract and process quantities of carbonic acid that considerably exceed its average content in the troposphere. In some cases it is possible to speak about the significance of carbonaceous fertilizers. The increase of organic matter as an effect of such "fertilization" is not in strict proportion to the increase of carbonic acid in the air.

The quantities of carbonic acid of the order of 1×10^{-1} % by weight, and even more, are generally favorable for life, while its average quantity in the troposphere is close to 3×10^{-2} %; that is, the quantities of carbonic acid easily used by green plants are five or even ten times greater than those existing. Nowhere in the Earth's crust do we see quantities of carbonic acid stifling green life; we can observe the green plants' innumerable adaptations to living with its relatively small quantities. The potential of living beings' power for extracting carbonic acid and processing it into matter necessary for life considerably exceeds the actual possibility of this transformation existing in the biosphere. On land, in the fields and in the forests, the picture of nature shows a still closer connection between carbonic acid and life. Organisms – terrestrial green plants – use mainly the carbonic acid of *biochemical origin* (i.e., that created by life itself) to build up their bodies, and thereby creating the biosphere's living matter.

De Saussure, at the transition from the eighteenth to the nineteenth century, already understood the significance for life of terrestrial green organisms and the carbonic acid of the atmosphere. He also paid attention to the carbonic acid released by the soil. At present, the latest studies confirm this significance on a scale that earlier scientists could not even have imagined. Lundegaard arrived at the apparently correct conclusion, based on his own and other scientists' experiments, that green plants process and absorb mainly the carbonic acid originating from the *respiration of the soil*, and that the carbonic acid of the troposphere only regulates the assimilation of carbonic acid by green plants, serving as a reservoir from which they draw it if necessary. "The respiration of the soil" is a biochemical process related to the carbonic acid released by bacteria, soil microfauna, and fungi. The intense proliferation of these organisms is closely connected with intense gas exchange, with the respiration (of CO_2) of a myriad of small organisms inhabiting a thin layer of soil, mainly 5–15 cm deep. This thin layer of 10 cm is the most decisive in the history of terrestrial life.

Life of the hydrosphere, however, presents quite a different picture. In the hydrosphere, the main mass of green life is concentrated. Apparently, this life fully depends on the troposphere; the carbonic acid of the air is its main source, and not only as a regulator of the absorption and assimilation of carbonic acid. It should be remembered that the carbonic acid regime in the ocean is not quite clear yet, and that our contemporary ideas about it will undergo great and numerous corrections. The thorough study of sea reservoirs points out a higher content of carbonic acid in deeper horizons, close to the bottom. This phenomenon could be caused mainly by the vital activity of bacteria (for instance in the Black Sea). That is why it is possible that the ocean's green organisms draw carbonic acid not only from the air, but also from the depths.

15   the dynamic balance of carbonic acid in the atmosphere

The quantity of carbonic acid in the troposphere is subject to continual oscillations, more so than the quantities of other gases comprising it. These oscillations however, do not reach the enormous range observed for water vapor. Still, they can reach very high figures in the layer accessible to terrestrial plants, usually dozens of percentage, and sometimes more than 100%. These oscillations are much smaller in the atmospheric layers that are far from the soil, and less stable. The constancy of its average content in these layers can be spoken about with more certainty than that of water vapor.

The mass of carbonic acid constantly present in the troposphere is close to 2.2×10^{12} t. Much greater quantities (1.84×10^{14} t) are present in seawater, partly in the form of a gas solution, and partly combined with bicarbonates. The mass of water in the ocean, as Schlesing has proven, is a powerful regulator of carbonic acid in the biosphere. Ocean water returns it to the air when the pressure of the atmospheric carbonic acid vapor decreases, and absorbs it again when the pressure increases. This important fact was discovered by T. Schlesing in 1878 and was confirmed later by independent studies. This fact, no doubt, is of great geological importance, for the ratio between the areas of ocean and land was very changeable throughout geological time. Its geological importance is not completely understood yet.

Oceans always contain carbonic acid in quantities required by life and favorable for it. Seawater is a little alkaline; it lacks a few hydrogen ions to be neutral. This lack of H^+ greatly affects the life of organisms, and the more deeply the ion state of water is studied, the clearer is its significance. L. Henderson (1914) is right in stating that the regulation of this ionization (i.e., of the number of hydrogen ions in water) takes place with an exactitude inaccessible to the modern means of our laboratories, and resembles the exactitude of celestial movements. This assertion is not exaggerated.

Oceans are powerful regulators, but the regime of the atmosphere, of its movement, is at the same time affected by two other phenomena of the Earth's crust: inland water surfaces and living matter. The mass of the latter, however, is not very great; it hardly exceeds $10^{-2}\%$ of the world ocean. But its influence cannot be ignored, for carbonic acid is part of the cycle of life, and while the ocean's surface releases an amount of carbonic acid that by no means can be characterized by its mass, and remains unchanged during the year, the surface of minute, rapidly proliferating organisms allows a release that can not only equal or become a value of the same order as the ocean's surface, but even exceed it. The biogenic process of carbon dioxide formation has no limit for the ocean, because of the resilience of the carbon dioxide in the atmosphere. But the existence of these regulators also affect life; they change, and maybe even lower the optimal conditions existing for an organism in life's environment, and reduce the quantity of living mass that could be created by proliferation in their absence.

There is a power in the Earth's crust equal to the geochemical energy of life, and this power balances its effect by combining with the carbonic acid of the air. This power is presented by the world ocean and the areas of fresh water. It manifests itself in the resilience of this CO_2 in terrestrial water solutions and in the formation of bicarbonates in these waters. This force prevents living matter from extracting the quantity of carbonic acid from the air or water that it is able to extract, and that it would consume when it is given abundant gaseous carbonic acid. Like any balancing phenomenon, the mass of living matter (as well as its proliferation, which is a function of carbonic acid absorbed by living matter) is restricted by another natural phenomenon, which is as powerful as itself.

The observed phenomenon is very complex, and we cannot consider it in all its aspects. Apparently, for terrestrial green plants, the regulation of its quantity can be measured only by the influence the ocean exerts on carbonic acid. The bulk of carbonic acid is produced by respiration of soil though, created by life itself – not by green matter – and does not depend on the troposphere. *Coming to land as green matter, life leaves the restricting tension exerted on it by the world ocean.*

But the ocean itself is not a simple water solution; it is filled with life. A complex balance unknown to us must exist between the quantity of marine living matter rich in animals, the chemical properties of bicarbonates, the mass of carbonic acid in seawater, the biochemical "respiration" of soil, and the mass of terrestrial green plants.

Thus, carbonic acid in the troposphere is in a state of complex dynamic balance.

To sum it up, we see that carbonic acid frees itself into the atmosphere:

1. By way of volcanic eruptions, hot mineral sources, natural gases, and by weathering of igneous rocks (juvenile or phreatic carbonic acid).
2. From solutions in seas and fresh waters (lakes and rivers) in connection with the resilience of the atmospheric carbonic acid (vadose, partly biogenic carbonic acid).
3. Through the respiration of plants and animals during their life, through chemical and biochemical processes related to their decomposition after death (their decay), and through soils (always only biogenic carbonic acid).

And finally, the carbonic acid that is frequently released in coal deposits of Western Europe is in all probability from the Quaternary. Part of it must be biogenic, and part of it is probably related to the gaseous isolation of underground atmospheres.

The release of carbonic acid by Man in the process of his technical work is considered biogenic, such as the release occurring in factory furnaces, calcinating lime, fermentation, and in many other processes. It is a very interesting and characteristic fact in the history of carbon that the quantity of carbonic acid released by mankind in this way increases with the progress of civilization. It has already reached such an order that it must be taken into account in the geochemical history of the biosphere.

Thus, according to A. Krogh's calculations, the quantity of carbonic acid released by the consumption of coal reached 7×10^8 tons in 1904, and rose to 1×10^9 tons in 1919 (F. Clarke). This amounts to as much as 0.05% of the entire mass of carbonic acid existing in the atmosphere. Such an increase acquires the status of an important geochemical phenomenon. In this way, civilized Man breaks the established terrestrial balance. With the civilization of *Homo sapiens*, a new geological power has appeared, the significance of which in the geochemical history of all elements is growing. We will see that it is only one fact of a great, general, historically new phenomenon.

Carbon is an element playing a great role in planetary space, as meteorites and comets show. No doubt, the burning of carbonaceous meteorites and shooting stars is a source of cosmic carbonic acid [i.e., CO_2] that is incessantly penetrating the Earth. A dynamic balance is being established every moment. The carbonic acid penetrating into the atmosphere in this or that way does not remain there; it passes on to neighboring areas. Its main part ends up on the Earth's surface – in the biosphere – in the form of soluble carbonates combining with metallic oxides, alkalines, and alkaline Earths that are related to silica.

As we have seen, silicates and aluminosilicates decompose on the Earth's surface, giving colloidal siliconic acid, silicon anhydride, and kaolin clays. The quantities of carbonic acid replacing silicon and aluminosilicon anhydrides are very large. Part of this acid, as carbonates or bicarbonates (ions CO_3^{2-} and HCO_3^{1-}) is captured by natural waters and is constantly carried to the ocean by rivers; this is a mass corresponding to about 1×10^9 t CO_2 per year. The whole quantity of carbonic acid taken up by the weathering of aluminosilicates and silicates (absorbed by them), is no less than 5×10^9 t, and probably more than 1×10^{10} t annually.

The smallest mass is consumed by living matter and is located in the bodies of living organisms. All the carbon of living organisms originates ultimately from the carbonic acid of the atmosphere; at some moment it certainly existed as gaseous CO_2. Acknowledging the data about the carbon of existing living matter, we can think that the total quantity of carbonic acid needed for its formation corresponds approximately to the order of 1–2×10^{13} t. This quantity is much larger than that of the atmospheric carbonic acid (2.2–2.4×10^{12} t). Usually it is considered to be smaller. For instance, Arrhenius assumes that the quantity of carbonic acid used annually by living matter does not exceed one fiftieth of the total mass of atmospheric carbonic acid. Arrhenius bases his conclusions on J. Liebig's data. Liebig estimated the average harvest from Central Europe's soils as 2.5 t of dry organic matter per hectare. Taking this figure as the average for the whole Earth's surface, Arrhenius arrives at 1.3×10^{10} t CO_2. It has already been mentioned that these figures do not correspond to reality.

It is not the figures given by the Swedish scientist I am criticizing here, but only his evaluation of the order of the phenomenon; this order is different. The quantity of carbonic acid retained in living matter at any moment is not smaller, but on the contrary, is at least dozens of times larger than its quantity in the Earth's atmosphere. Living matter is constantly losing [releasing] its carbon; it isolates it mainly in the course of vital processes as biogenic carbonic acid. It is obvious that in order to keep the balance of its composition, living matter must annually extract much larger carbon masses from the atmosphere than it contains on average. If this is true, the consequence is that all carbonic acid of the troposphere must pass through living matter many times during one year.[49]

The carbonic acid penetrates into living matter with the help of a special mechanism – the uncountable chlorophyll plastids dispersed in green plants. This apparatus allows a molecule of carbonic anhydride to decompose and to transfer carbon into carbohydrates or other, organogenic compounds. The greatest mass of carbonic acid is captured not by the terrestrial green plants we see around us, but by the unicellular algae of ocean plankton.[50] Both in the ocean and on land, biogenic absorption, isolation, and synthesis of carbonic acid take place in the biosphere. This occurs mainly in special areas where the organisms are concentrated – *in films and concentrations of life* that are unchangeable and geologically eternal, notwithstanding the sharp variety of life forms they consist of.

There is, finally, a third source of absorption and release of carbonic acid in relation to the atmosphere. This is the ocean and the surface of other waters, whose great role has already been mentioned. Thus, it

---

49. Modern estimates suggest that the CO_2 in the atmosphere is about 6.2 to 7.5 times the amount of CO_2 that is absorbed via photosynthesis each year. Thus, if only "unused" CO_2 were absorbed by green plants, it would take 6.2 to 7.5 years to "use up" all the CO_2 of the atmosphere; on the order of 1/10 of the CO_2 in the atmosphere enters and passes through the biosphere each year – which is still a very impressive amount! –E*d*.

50. In Vernadsky's time it was thought that the oceans accounted for twice as much photosynthesis as the land (because ocean area is about double that of the land). It is now thought that the numbers are reversed: that land organisms account for about twice as much photosynthesis as those in the ocean. This is because vast areas of the oceans have been found to be nutrient deserts: They have too few mineral nutrients (especially Fe) to support much life. Nutrients are absorbed by organisms that may die and take the nutrients to the ocean bottoms. In some areas (e.g., off the coast of Peru), currents from the depths bring up nutrients, which support flourishing ocean life. We will encounter this idea again in *The Biosphere*. –E*d*.

is inevitable that the surface water masses must control the quantity of atmospheric carbonic acid. The balance is constantly being established. A break of this balance, had it existed, would have an immediate effect on the whole chemistry of the ocean. But the chemistry of the sea looks unchangeable to us. Throughout geological time, we always see the same reactions and minerals as we see today. The same "life conglomerations" have always been present in the ocean.

The stability of the chemical composition and properties of marine sediments shows that, throughout geological time, there have not been great and sudden changes in the quantity of atmospheric carbonic acid. There could be, and there were, only its oscillations. The average of this quantity has always remained close to the present-day value; a dynamic balance of carbonic acid had to exist. In the geological past, we don't see any signs of its change in some definite direction – neither of its increase, nor of its decrease. Oscillations in the CO_2 content throughout geological time seem theoretically inevitable, but there are no clear geological indications of it. Arrhenius saw them in the repeated appearance of glacial periods, which he explained by a change of the thermal transparency of the atmosphere, caused by a different content of carbonic acid in it. But the causes of glacial periods are much more complex, and the oscillations in the content of carbonic acid cannot account for it.

Acknowledging the existence of centenary or even geologically long oscillations, we can assume that the quantity of carbonic acid in the atmosphere does not remain absolutely stable at present either. Arrhenius, on completing his studies, expresses the opinion that its quantity at present is constantly increasing. He has pointed out a new fact in its history, which was absent in the previous geological eras – the activities of civilized Man. We have already seen the significance of these activities in the release of carbonic acid. But the human activities have not yet been summed up, and it is clear that the human influence concerns not only the release, but also the absorption of carbonic acid, connected, for instance, with the change of green plant matter in the biosphere caused by Man. The agricultural areas are on the whole incomparable to the natural life conglomerates they have replaced. The increase in green living matter as a result of the agricultural work of Man must inevitably absorb the carbonic acid created by the human technical activities.[51] Maybe the dynamic balance, so characteristic of natural phenomena, is in an unconscious, "spontaneous" way for humans as well.

16   the cycle of life

I will call the balance between carbonic acid and living matter a *cycle of life*.[52] It is of enormous significance in the geochemistry

---

51. Based on the situation today, it appears that Arrhenius was correct, and Vernadsky was over optimistic. The absorption of CO_2 by green plants and by the oceans does increase as the CO_2 concentration in the atmosphere increases, but the great increase in burning of fossil fuels since Vernadsky's time plus the clearing and burning of forests and the usual sources of CO_2 (respiration, decay, volcanoes, etc.) seem to be winning out: In the 1700s to 1800s, the concentration was about 270 micromoles of CO_2 per mole of air (equivalent to parts per million); by the 1930s (Vernadsky's time), it had reached about 300 μmol/mol; and by the year 2000, it was about 370 μmol/mol. Thus, the increase from 1700 to 1930 was on the order of 10%, but from 1930 to 2000, the increase was about 23%. –*Ed.*

52. Here, Vernadsky defines what we generally call "the carbon cycle" as the "life cycle." But in modern biology, "life cycle" refers to the development of an organism from a fertilized egg (zygote) to the production of another zygote, which begins the next generation. Hence, to retain but modify Vernadsky's "life cycle," I have changed "life cycle" to "the cycle of life" – meaning "the carbon cycle." –E*d.*

of carbon. The most characteristic feature of the cycle of life is its incomplete reversibility, as it returns part of the carbonic acid absorbed by life to the environment. Some of its atoms are always retained in the cycle of life, and another part is isolated as biogenic carbon minerals. The carbon of the latter leaves the geochemical cycle and returns there sometimes after a geologically long period of time. The principal groups of such biogenic carbon minerals are lime carbonates, coal, oil, and bitumen. All the others originate from them or make up insignificant masses compared to them. The history of carbonic acid – the cycle of life – can thus be represented by the following scheme:

table 12   the cycle of life

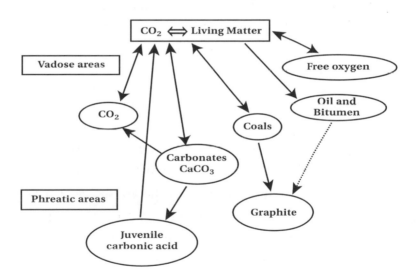

The quantities of carbon lost from the cycle of life as carbonates, coal, and oil amount to only an insignificant part of the total organic carbon. Living matter keeps carbon in the cycle of life. This is the most characteristic feature of this element's geochemistry. The bulk of the carbonic acid absorbed by organisms is always retained by living matter. Even when the carbonic acid is released by some of these organisms, it is immediately seized by others (the respiration of soil).

This fact was described by the great natural scientist, K. M. von Baer, in 1838 as the *law of nature's economy regarding living matter*. K. M. von Baer noticed that the transformation of organic carbon into non-organic carbon (i.e., the creation of vadose carbonaceous minerals) develops with an extreme slowness. Carbon leaves the cycle of life only in very small quantities. But still, it partly leaves the cycle, and thus the cycle becomes irreversible. This fact is of extreme significance for Earth's history.

17   incomplete reversibility of the cycle of life

It is interesting that the importance of this phenomenon and its regularity was noticed still earlier and described in the outstanding, now forgotten works of the Polish scientist and physician, J. Sniadecky, in the eighteenth century. Sniadecky thought that this carbon returns to the cycle of life again due to various geochemical processes, for instance via volcanoes. This is true if volcanic phenomena are interpreted broadly, taking into consideration the magmas of phreatic envelopes, burning organic matter, and decomposing carbonates.

Disappearance of carbon from the cycle of life is not accidental. Lime carbonates, coal, bitumen, and crude oil have originated from living matter. Certain stable chemical compounds can always be found in organisms, usually in minimal quantities. They form during the life of the organism, and after its death and decomposition they serve as a basis for carbonaceous vadose and phreatic minerals. Only a small part of carbonaceous compounds passes to the minerals; that is, leaves the cycle of life for a long or a short period of time. The complex mechanism of this change has only lately begun to be discovered in its totality.

Limestone is, for example, formed mainly from shells and in general from skeleton parts of organisms, the composition of which corresponds to the close relation between calcium carbonate and special nitrous organic compounds. Even if calcium carbonate forms in a different way, its synthesis in nature is mostly biogenic; it is accumulated in the warm waters of oceans, seas, and salty lakes by unicellular algae such as oolites, or is concentrated in great masses under the influence of bacteria (such as *Bacillus calcis*) and fungi (*Actinomycetes*). In 1903, G. A. Nadson discovered this bacterial process of great importance[53] for seas and salty lakes. It had been forgotten, and only in 1914 was it rediscovered for oceans and seas by the young scientist Drew, whose death tore him away from science so early. The deeper this phenomenon is studied, the more important it turns out to be, and the more it begins to manifest itself to us as a part of a more complex biogenic process.

On the whole, the biogenic process of the creation of limestone accumulations is not restricted to the isolation of the organically rich lime [calcium] parts of organisms and their mechanical aggregation after the aquatic organisms' death, or to the biogenic aquatic

---

53. G.Nadson, *Microorganisms as Geological Actors,*p. 60, 71, P., 1903

isolation of solid calcium carbonate by microbes. It also involves the activities of boring algae that decompose, and precipitate or cement, the loose, mechanically gathered mass of lime remains again. This process, which appears to be very important for the migration of calcium, is just beginning to reveal itself to us.

There is another process that deserves attention, but that has not yet been comprehended and still arouses doubt: the sedimentation of calcium carbonate in the seas under the influence of the decomposition products of organisms (bacteria?). In any case, bacteria and other microbes create a great portion of the limestone. Their role in this process is probably passive, since the sedimentation of calcium ions from the environment takes place under the influence of the CO_2 released in all life processes. The vital importance of microbial life is just being discovered for coal as well. I have already mentioned the geological conditions of its formation, and its relation to certain organisms. The dominant part of coal – the humus – is genetically related to lignin, which forms in a living organism during its life, and which has a definite chemical structure.

In the creation of the humus compounds of coal, a major role is played by microorganisms that relate in different ways to the carbonaceous compounds making up the plant's body. These organisms destroy some of the compounds by turning them into CO_2 and H_2O but do not touch others. At first, R. Rose and Lieske (1917) showed the role of fungi in this process, and then F. Fischer and Schrader (1921) showed that of bacteria. The initial product of coal is not cellulose as had long been thought, but molecules of an aromatic structure – lignins. The quantity of lignin increases as the plants turn into peat and lignite, because of the process of their decomposition by microbes; aerobic bacteria and fungi destroy both lignin and cellulose, whereas anaerobic ones do not, or hardly, decompose lignins (S. Waxman, 1927).

The same phenomenon is observed for crude oil: their oxygen and nitrogen compounds – methylquinolines – and probably also cholesterol derivatives, are made by *living* organisms. This takes place during their life, before they get temporarily lost from the cycle of life. The history of *hydrocarbonates* and their parts are still unknown to us, but their basis is created by living matter. Lignins of coal, and methylquinolines of crude oil, are insignificant in mass when they are still part of an organism, but eventually they gather in great quantities in vadose or phreatic carbon minerals in the form of derivatives.

Also, they are preserved only under special conditions; usually they do not leave the cycle of life. Organisms have many devices for their destruction, for directing their chemical elements into the cycle of life. Calcium carbonate – like lignin – is destroyed by many special organisms that need it for life. Thus, the bulk of carbon atoms of living matter always returns either to living matter or to carbonic acid. A small part of this mass concentrates in vadose minerals such as calcium carbonate, humus, or resinous organic matter. These new products disperse everywhere, permeating all matter of the Earth.

Only in some cases they form agglomerates and concentrations of carbon, such as deposits of coal, resinous shale, asphalt, crude oil, and limestone. In general, terrestrial living matter – higher plants and animals – does not provide considerable accumulations of vadose carbon minerals. Soil humus and guano of tropical countries make up the largest accumulations that have appeared as a result of the decomposition of living matter's remains on the Earth's surface.

If these vadose minerals and their mixtures suffer further changes in deeper layers of the Earth's crust, the quantity of carbonaceous phreatic minerals is always very small in mass.

Carbon leaves terrestrial life mainly as dust; powder that manifests itself in dispersions. This dust can give accumulations of new carbonaceous minerals only in deep juvenile and phreatic areas. In those areas the dispersed matter is unstable; its recrystallization and agglomeration takes place, and at last it oxidizes into CO_2. Large accumulations of CO_2 that are lost from the cycle of life form under certain conditions in the aquatic medium. This is accompanied by sedimentation of products in inland aquatic reservoirs, from which coal, bitumen, and crude oil forms. Sometimes they are connected with the sea's outskirts. As for seas, accumulations of carbonates mainly form there in a different, but still biogenic way.

During geological time, seas and land change places: Sea appears where land was before and vice versa. In the areas where carbon has migrated biogenically in the form of calcium carbonate, it begins to accumulate in the form of organic minerals, for example. Only the oceanic spaces are outside the realm of these phenomena; there are no accumulations of organogenic carbon compounds there. Thus we see that the accumulations of carbonaceous minerals are part of a vast geographical and physical process related to geological time and to the cycle of life of living nature.

18   the geological cycle of calcium carbonate

It is obvious that the larger part of the carbon capable of leaving the cycle '$CO_2 \Leftrightarrow$ living matter' is removed from it as calcium carbonate, since the greatest mass of living matter is concentrated in seas, inseparably connected with the world ocean. Carbonates dominate among vadose carbonaceous minerals, and among the carbonates, calcium carbonate dominates. The formation of carbonates is connected with the history of calcium, the quantity of which exceeds the quantity of carbon in the Earth's crust by several times. If the quantity of carbon in the Earth's crust corresponds to 1.3×10^{17} t, that of calcium approximately equals 1.1×10^{18} t.

These figures, which apply to the whole crust, present nothing but the limit values for life's direct environment – the biosphere – for at present the ratio of Ca:C (in mass) in the biosphere cannot be numerically calculated. We can only postulate that its value equals 7.8 in the Earth's crust and decreases sharply in the biosphere, where it hardly reaches one-third (hardly exceeds 3.0). In the biosphere, all the conditions are established for the biogenic creation of calcium carbonaceous compounds, as both these elements are concentrated in living matter.

Vadose carbon minerals exist everywhere in the biosphere, and sometimes, as in the case of limestone or coal, in enormous quantities. The illusion even appears that they are more important in the geological history of calcium and carbon than they are in reality, for it takes an effort to remember that they are nothing but small remains of the mass of extinct living matter that had once created them. The quantity of carbon annually leaving the cycle of life is small, but during geological time it gathers into vast masses. table 13 presents in decreasing order the quantities of CO_2 corresponding to the carbon of mainly biogenic origin existing in the biosphere and the stratisphere.

table 13

Carbon (CO_2)	Amount
1. Limestone(CO_2)	3.1×10^{16}t
2. Living matter (CO_2)	5×10^{13}–5×10^{14} t
3. Ocean (CO_2)	1.0×10^{14} t
4. Coal (CO_2 according to the calculations of the International Geological Congress)	2.2×10^{13} t
5. Atmosphere (CO_2)	2.2×10^{12} t

These figures show the great mass of carbonic acid that has passed from organisms to limestone over geological time, and which has not proceeded back into the geochemical migration during that time. Apparently, we are dealing only with a small part of the carbonic acid that has ever existed as $CaCO_3$. Nevertheless, these figures can give an idea of the order of the phenomenon. In all probability, *the quantity of CO_2 freed from the cycle of life as calcium carbonate is larger by hundreds of times than the quantity of carbonic acid existing at the present moment in the atmosphere, the ocean, living matter, and the technically accessible coal layers* (taking into account the content of carbon).

It still requires checking, but the carbonic acid of this origin ($CaCO_3$) will remain larger in mass even if we add to the above mentioned sources the yearly balance of carbonic acid released into the biosphere by gas streams, mineral springs, and volcanic eruptions. At the same time, everything shows that the mass of organogenic limestone and other organogenic concentrations of calcium carbonate in the biosphere, stratisphere, and upper metamorphic geosphere, is *smaller* than the mass of the carbonic acid corresponding to the *dispersed* carbonaceous compounds, bitumen, hydrocarbonate, and elementary carbon of different forms. All matter on Earth is permeated with this carbon, and there are no vadose or phreatic rocks that lack it. Its quantity can be smaller in juvenile rocks, but still it is larger than we can calculate now.

A few figures can give an idea about the order of these quantities. Thus, rivers annually pour organic matter into the ocean in quantities exceeding 2×10^8 t. The quantity of this matter dissolved in seawater corresponds to a mass of at least 5×10^{12} t, and probably even more. But water accounts for no more than 7% of the Earth's crust (the region up to 16 km in thickness), and the same traces are formed everywhere in much larger quantities. Their enormous quantities are dispersed in sedimentary rocks, except for limestone and dolomite. According to F. Clarke, no less than 7×10^{15} t of carbon exists in such a state. But F. Clarke took into account only the shale and clay rocks, leaving out sandstone; we know that there is no sedimentary quartz without organic matter. The figure of 7×10^{15} t of carbon should be increased at least with several units to $n \times 10^{16}$ t of organic and other matter containing carbon.

This carbon dispersion is connected with every chemical phenomenon, every life process. I have mentioned the large quantity of carbonic acid annually released by Man into the atmosphere by burning various kinds of igneous matter. Before the war, this quantity exceeded 1×10^9 t CO_2. According to A. Smith's experiments – though old ones – at least 1×10^7 tons a year were dispersed in 1871 as thin bitumen and soot. This figure must at present be increased by several hundred times, taking into account both the growth of our technical civilization and the fact that Smith considered only some of the civilized countries of his time.

So, two facts can be considered to be proven:

1. The bulk of the carbon mass freeing itself from the cycle of life exists in a state of dispersion in all of the Earth's matter.
2. The largest part of this carbon exists there as carbonates, and mainly as calcium carbonate.

This carbonate formation – in this form and in the form of limestone – is of great importance in the geochemical history of calcium. I cannot dwell upon its history in detail here. I shall only say a few words about the enormous biogenic importance of this phenomenon. We hardly find any non-organic limestone – chemically pure sediments of $CaCO_3$. An attentive study of the terrestrial reactions shows more and more distinctly that these sediments are mainly biochemical.

At present we don't know any vadose chemical reaction giving large accumulations of calcium carbonate that are independent from life. No doubt, there are cases when calcium carbonate is formed solely by chemical reactions, but in those cases it is dispersed and makes up small, scattered concentrations that are almost immediately absorbed by organisms again. As we have seen, calcium is one of the most widespread metals in organisms. Its average quantity in living matter must exceed 1% of the mass; this figure is of the same order as that of carbon and nitrogen. Organisms seek it everywhere; if they find it amongst calcium dispersions that are accessible to them, they absorb it immediately, as for instance from dust, which always contains it.

Due to this, calcium carbonate in a dispersed state immediately enters living matter – the cycle of life. An organism does a tremendous amount of work to get it. For example, according to K.

Bischof's calculations, one oyster, in order to build its shell, must pass a quantity of water through its body exceeding its weight by $2.7 \times 10^4 - 6.6 \times 10^4$ times. This work is being performed now as it has been performed throughout geological time, mainly in the ocean, by an innumerable quantity of organisms.

In marine water, calcium carbonate does not exist as such. Calcium is introduced to it by rivers as carbonates and bicarbonates, mainly separated into ions. In seawater its quantity decreases very quickly under the influence of life, and its existence there as carbonate is improbable. Calcium carbonate forms in a solid state almost exclusively in living matter, or in a biochemical way. The formation of calcium carbonate in the surface parts of the sea was often pointed out, but it is becoming more and more obvious that this is also a biochemical process related to microbial life.

In organisms, calcium carbonate is crystallized in tissues in obvious connection with physiological functions. It is always closely interwoven with nitrous organic matter. With a lack of life in natural conditions and a rapid vaporization of salty water, calcium is concentrated mainly as sulfates. The biogenic isolation of calcium in seawater is performed only by carbonates. In seawater full of life, the sedimentation of carbonates occurs in a biochemical way; sulfates begin to form only in concentrated waters already devoid of life. This phenomenon does not depend on the progression of geological time; it has existed at least since the Cambrian era.

But at present it is far from being clear whether it is possible to go any further, and to claim that it has been so throughout geological history. Now we are facing one of the greatest problems of geology: that of the sudden appearance of highly developed life at the beginning of the Cambrian era, about a billion years ago. According to our current ideas, it should be supposed that hundreds of millions of years had passed before these forms of living matter were established, but their remains are absent from older layers. This cannot be explained by insufficient geological knowledge, for in general, geological studies have covered the whole Earth. It is even possible to make a correction and to add to the Cambrian the Algonquian era, where some calcium organisms resembling the Cambrian ones have been found. We shall descend several hundred million years more, and go past $1.1–1.2 \times 10^9$ years. Still the fact will remain; beyond the Cambrian (probably upper Algonquian) there are no calcium organisms (i.e., those having a calcium skeleton).

Thus, it must be acknowledged that either:

1. About that time a sharp change took place in the history of calcium in the Earth's crust, and the first organisms with a calcium skeleton appeared.
2. Or that this is a misleading phenomenon, and a result of the fact that in metamorphism phenomena, the inevitable processes of dissolution accumulate in the course of time, and the calcium skeletons of all organisms cannot be preserved. But this idea is hardly probable, as the total metamorphism of many Cambrian accumulations is very small, and does not comply with this assumption.

Similarly, it cannot be assumed that the calcium function of living matter before the Algonquian had not existed in one form or another, for no change has been observed in the character of the weathering of the Cambrian accumulations. Besides, in the Precambrian accumulations there are thick layers of lime that in no way differ from ordinary organogenic limestone and rocks, which have metamorphically originated from them (such as the Granville series in Canada). There are no grounds – taking into account the special importance of calcium in life phenomena – on which to attach some special origin to these limestones.

Thus, if nothing unexpected happens, it must be assumed that *the geochemical function of calcium more or less radically changed its morphological structure at least as late as in the upper Algonquian or even in pre-Algonquian time*. There is no need for ideas allowing a geologically rapid morphological change of the extremely complex Algonquian living realm – a mass creation of organisms containing skeletons – or a radical change of the ocean's composition – a geochemical catastrophe – if there are no other signs of it besides the lack of remains of calcium organisms in ancient layers. No negative data relating to those periods have been found regarding one of the main manifestations of calcium's biogeochemical function that takes place at present in the crust itself in the form of a bacterial process – the creation of calcium carbonate. It could have taken place continually at that time, too.

Bearing this in mind, it is possible to assert that in the ocean, which had in general the same composition as it has now, innumerable species of organisms concentrating calcium – *calcium organisms and those rich in calcium* – existed throughout geological time. The first contain more than 10% of calcium in their living mass. Calcium is contained in them in a larger quantity than carbon, and its percentage is of the same order as the content of oxygen and hydrogen. These organisms, beginning with Algonquian mollusks, brachiopoda, echinodermata, corals, hydroids, worms, algae, crinoideae, rhizopoda, bacteria, etc., leave vast quantities of carbon and calcium in the form of carbonates after they die. A part of these carbonates dissolves again and immediately enters living matter of the cycle of life. Another part is absorbed by other organisms and does not leave the cycle of life anymore. But a very considerable number of shells and skeleton parts or organic accumulations of carbonates (bacteria and their products) remains more or less untouched and forms calcium and carbon accumulations. When, due to geological phenomena, such as the accumulation of new sediments, this is isolated from the influence of living matter, it gives limestone upon further diagenesis.

Now it is obvious that the bulk of calcium carbonate accumulations is not formed by geologically more recent and larger organisms drawing our attention within the vast richness of ocean life. The largest part is formed by organisms invisible to the eye, which have existed ever since the Archeozoic era. For a long time, only foraminifera were taken into account. Their importance in the formation of limestone was proved almost a hundred years ago by Ehrenberg, who had introduced into science Linnaeus's old principle "omne calx e vermibus" – all the lime originates from worms (invertebrates). About 20 years ago, it was discovered that incomparably smaller beings are still more important, such as the planktonic unicellular algae called coccolythophorides, coccospheres, and rabdospheres. They exist in enormous quantities and proliferate with an incredible speed.

The growth of calciferous chlorophyll organisms is connected with the speed of their chlorophyll synthesis, and this accounts for the importance of these organisms, such as for instance zooxanthella living in the polyps of corals. In the twentieth century, new organisms were discovered – even smaller calciferous bacteria – which apparently present a still more powerful mechanism for the formation of calcium carbonate. It is possible to measure the quantity, or to be more precise, the order of the quantity of calcium carbonate sedimented by life on the ocean bottom. It is enormous; it can be said that all the calcium annually poured into the ocean by rivers in the form of carbonates and bicarbonates accumulates there at the same time as biogenic calcium carbonate. Clarke's numbers show this clearly in table 14:

table 14

Calcium	Amount
The quantity of calcium annually poured into the ocean in a dissolved state, combined with CO_3^{-2} and HCO_3^- ions	4.56×10^8 t
The quantity of calcium annually sedimented in the ocean as limestone	6.6×10^8 t

Thus, the secondary geochemical cycle of carbon on the Earth's surface is a partly irreversible cycle of life. Some of the carbon is constantly leaving it. The quantity of this disappearing carbon, although it seems so large to us, presents nothing but an insignificant part of all carbon in the cycle of life: $CO_2 \rightleftarrows$ living matter. The bulk of this carbon, as we have seen, accumulates as $CaCO_3$. Annually, 2.2×10^8 t of carbon are accumulated in the ocean in this way. At first sight it may seem that this is the larger part of the carbon existing in living matter. But the carbon mass that annually passes through living matter to the ocean is larger than the mass contained in living matter at any given moment. We know that it is vast, but cannot yet express it numerically.

We must compare the carbon of calcium carbonate with the whole mass of carbon related to the cycle of life, and not just with the carbon of living matter. It is evident that in this case it will correspond to fractions of a percent of all the carbon involved in the cycle of life. But these small fractions of a percent, gathered throughout geological centuries, give an enormous amount of matter, and the process of their formation is of paramount importance in the chemistry of the Earth's crust. Apparently, this mass is limited and does not grow infinitely throughout geological time. As a result, its carbon changes to gaseous carbonic acid again, and returns to the cycle of life.

It can be supposed (and naturalists like J. Sniadecky already thought so) that all the carbon involved in the cycle of life sooner or later returns there from the Earth's depths (i.e., from the phreatic area). But we cannot prove this now. We must even assume that if it does return to the cycle of life, it takes place at the expense of energy that is independent from the solar radiation sustaining the cycle of life. Under the influence of ozone, and most probably of oxygen, only a small part of the dispersed carbon gradually turns into carbonic acid on the Earth's surface, and thus is taken up through the cycle of life again. The main part of such carbon turns into CO_2 only in the depths – in the metamorphic and juvenile areas – because of the heat of the deep layers of the Earth's crust. As long as carbon does not travel that far, it remains outside the cycle of life. The dispersed carbon can remain in such a state for an indefinite amount of time.

Sometimes it remains outside the cycle of life for geological eras in the form of combined carbon, such as calcium carbonate, coal, crude oil, and bitumen. After leaving the cycle, carbon creates, and supports by its exit, phenomena of the greatest significance in the biosphere, *for only in this way is a possibility created for the existence of the corresponding masses of free oxygen in the biosphere.* In the life processes, living matter creates oxygen in a free state from carbonic acid and water. All the oxygen of the atmosphere and water solutions is of such origin. If carbon did not leave the cycle of life as hydrocarbonate, coal, bitumen, graphite, or as calcium carbonate, free oxygen would not exist at all; as a result, thousands of vital chemical reactions of the biosphere would not exist either, for the free oxygen of the Earth's crust is very active: It possesses free chemical energy to a degree incomparable with other elements. Hence, a simple correlation must exist – still unknown to us – between the quantity of free oxygen of our planet – of its biosphere – and the mass of coal, bitumen, oil, and carbonate existing in it.

Thus, the study of the irreversibility of the cycle of life leads us to even deeper problems. It leads us to a scientific area that is still in the process of formation and that concerns great tasks of life and energetics. This science is the realm of the future – *the future energetics of our planet*. About 90 years ago, the German physician, J. R. von Mayer, first understood that green plants change the energetics of the Earth's crust by the very fact of their existence. They turn the radiant energy of the Sun into a new form that is favorable for the chemical processes of our planet. These ideas of J. R. von Mayer were not understood long after his ideas about the unity and preservation of energy had penetrated scientific thought. Von Mayer returned to these ideas several times after their first publication. They led him to the concept (since then taken for granted) that coal contains potential energy – the energy of solar rays belonging to previous geological eras – and that Man, by using these fossils, uses this fossil energy again.

Living matter in the form of the realm of green plants accumulates solar energy. The power gathered in this way can be preserved for millions of years as coal – a carbon mineral of vadose origin. Now we must generalize von Mayers' idea: *Through living matter, solar energy exists in a potential state not only in coal originating directly from green plants, but in all vadose minerals of carbon, such as in calcium carbonate and other biogenic minerals – in the majority of vadose minerals, and I think in all of them to a great measure.*[54]

No doubt, all the chemical combinations connected with life are gatherers of solar energy. Even if energy manifests itself in them as molecular and chemical energy, the existence of the compounds has become possible only thanks to the radiant energy of the Sun, captured by a living organism and turned into chemical energy. Here we touch upon the deepest phenomena of life studied by science at any time. These phenomena cannot be ignored if we want to understand the philosophical and scientific importance of the mentioned ideas and all the conclusions hidden in them.

19   redi's principle

From this standpoint, two general phenomena in the course of life on the Earth's surface draw our attention. First, the existence of a distinct border between living and inert matter, and second, the

---

54. The idea that coal is the fossil energy of solar rays that had passed through living matter arouses no objections and seems clear, but that this is the case for any phosphorite or biogenic mineral such as $CaCO_3$ usually meets objections. However, in both cases the phenomenon is the same: Solar energy creates through living matter a substance that cannot form outside the realm of life and outside the biosphere, and that sooner or later passes into another compound that is stable outside the life environment. The energy created by them is chemical, according to the form of a chemical compound such as lignin, $CaCO_3$, or phosphorite; it is created by life and only due to this a certain quantity of chemical energy is isolated upon their decomposition. I hope to dwell upon this issue in more detail in another paper.

special character of the energy associated with the manifestation of life. This energy seems different from the energy of almost all other natural processes.

Remaining in the realm of empirical facts, we can state that nowhere and never on our planet has *new* life been created without being materially connected with other life. In geochemical phenomena studied it has always been like this. If some faraway cosmic period existed that has not left traces in the geological history, in the "stones" of the planet, they are not subject to scientific study by geology and geochemistry. We must always distinguish positive biological facts from inevitable hypothetical cosmogonic presumptions, even if the latter are put into a scientific form. I do not doubt their usefulness for the progress of science, but in their exactitude and importance they are absolutely incomparable with the facts of observation and experience. One cannot base science on cosmogonic conclusions when there are no exact corresponding empirical facts, confirming with certainty the cosmogonic conclusions or causing them.

I will not touch upon the issue of eternity or the beginning of life in general; I dwelt upon the history and the state of this issue in a different place, and I have no reason to change my viewpoint. I will not touch upon the conditions of the appearance of life on our planet either. But it is necessary to make one essential remark: From the geological and geochemical points of view, the issue concerns not the synthesis of a separate organism, but the origin of the biosphere. The problem of abiogenesis, of the creation of a homunculus, cannot interest a geochemist; the only thing that can interest him and be important is the creation of complex life in the biosphere; that is, the creation of the biosphere. Does abiogenesis exist in nature or not? Has it existed in geological time? To answer this question, it is necessary to discover the form of life passing from generation to generation, which has provided its existence in the course of geological time (a phenomenon observed only in the biosphere).

More than 265 years have passed since the seventeenth century, when the Florentine scientist, physician, poet, and naturalist, F. Redi (1626–1697), first put forward an absolutely new idea for humankind. Several decades later – in the eighteenth century – it was generalized and deepened by another outstanding Italian naturalist, A. Vallisnieri. In the nineteenth century, Ocken followed Vallisnieri's thought and presented it as an aphorism: "Omne vivum e vivo" [Everything alive from the living]. This was a negation of the spontaneous generation and abiogenesis theory, and a proclamation of the continuous unity of living matter in our environment – the biosphere – from its very beginning, if that existed. After L. Pasteur's work, it is very difficult to shatter this view of nature, this empirical principle, which is based on many scientific facts. Although there still are attempts to prove the existence of abiogenesis, they are of no avail.

These centuries-long attempts are caused not by empirical facts, but by the habits of philosophical thought, by very deep traditions, which make up the basis of the ideas of the world, and which are associated with philosophical, religious, and poetic notions that are alien to science. Studying the geochemical history of carbon, we have not seen any traces of abiogenesis there; no organic compounds exist that were independent from living matter and that testified to the existence of such a process throughout geological time. Geochemistry proves the close connection of living matter with the history of all chemical elements and presents it to us as part of the organization of the Earth's crust, which is quite different from inert matter. In its data, there is no place for abiogenesis (spontaneous generation), since there are no signs of its existence.

We must retain Redi's empirical principle and acknowledge it as a still unshattered scientific fact, that throughout geological time, an insurmountable border has existed between living matter – the totality of all organisms – and inert matter. All life originates from life, and throughout geological time the phenomena of chemical exchange between these two manifestations of nature were the same as those observed at present. Within the limits of these empirical facts, the idea of the *eternity* of life seems quite natural. This idea permeates the religious and philosophical life of Asia and is beginning to penetrate into scientific ideas and philosophical searches of the West.

Living matter has always, throughout geological time, remained an inalienable, regular part of the biosphere. The source of energy captured by it from the solar rays – the matter – is in an active state, exerting a decisive influence on the course and direction of all geochemical processes in the Earth's crust. The inert matter of the Earth has been nothing of this kind throughout billions of years.

20 the energy of living matter and carnot's principle

Thus we are coming to the second empirical generalization characteristic in the general aspect of nature: the life phenomena – the special character of the energy of these processes. The history of ideas concerning the energetics of life taken within the limits of the cosmos shows an almost continuous succession of thinkers, scientists, and philosophers who came to the same ideas more or less independently, but who did not deepen the problems they had posed. The favorable atmosphere of present-day ideas seems to have been established long ago. We find short but clear thoughts, facts, and speculations about the energy difference between living and inert matter in the works of the founders of thermodynamics (J. R. von Mayer, W. Thomson (Lord Kelvin), and H. Helmholz).

These ideas were not understood and not appreciated. Later, the prematurely deceased S. A. Podolinsky understood their importance and tried to apply them to the study of economic phenomena. They play a significant role in the concepts of philosophers, especially in the philosophy of H. Bergson. But I think that the Dublin professor, J. Joly, was the first to see that the special energy character of *living matter* – the totality of living organisms – distinguishes living matter from *inert matter*, and several times he reached an important conclusion based on this idea. Since then this subject has often been revisited in the twentieth century. These ideas are penetrating our science deeper and deeper, although they have not yet acquired the necessary stability to take root in our notions of the world.

The geochemical history of carbon, most closely connected with living matter, leads to a different energy aspect of biogeochemical phenomena as compared with the geochemical phenomena taking place without life's influence. Natural phenomena expressed in terms of energetics are usually reduced to Carnot's principle. We know that they are always connected with degradation of energy; the quantity of free energy capable of doing work is reduced with every natural phenomenon.[55] The energy disperses as heat, and the world's entropy, as Clausius put it, increases and thereby levels the

---

55. This is part of the often-discussed *second law of thermodynamics. –Ed.*

heat. If the world has a limit, if the totality of natural phenomena is finite, the end of the world must follow – the leveling of energy that will not allow any natural phenomenon associated with energy to manifest itself. For a long time these conclusions were considered true consequences – laws of nature. No exception was found to this rule that had led to significant scientific discoveries.

The philosophical idea about the end of the world corresponded at the same time to deep human inclinations, to the ideal anthropomorphic notions of the world. Until now, the value and the general character of Carnot's principle is esteemed by scientists and philosophers in different ways. The evolution of contemporary scientific thought brings us to specify this esteem. A new, seemingly free field of mathematical and philosophical speculations opens in front of us, for our idea of the *correlation between matter and energy* is changing. Not only has our understanding of matter radically changed, the very notion of matter is beginning to change under the influence of empirical generalizations; the course and the effects of this change cannot even be imagined to a sufficient extent. Carnot's principle will inevitably find a new understanding. The manifestations of life are an empirical fact that is difficult to place within the limits of other natural phenomena in respect to Carnot's principle. The decrease of energy, its dispersion as heat, does not take place in the life of green chlorophyll plants (as we understand it), or autotrophic microbes seen in connection with the biosphere.

On the contrary, due to the fact of these organisms' existence, the quantity of free energy capable of doing work evidently increases toward the end of their life in the surrounding nature, and eventually in the course of geological time. The free oxygen produced by green plants, the coal forming from their remains, the organic compounds of their bodies, which nurture animals, and the movements and other chemical and physical manifestations, present new kinds of energy activities that are by no means accompanied by the degradation of the initial solar energy. This energy has passed to form, creating an organism that possesses potential immortality and which increases and does not decrease the active energy of the initial solar rays. Physiologists studying an individual animal organism outside its environment do not think it necessary to draw these conclusions. But the realm of animals exists only at the expense of green living matter and cannot exist separately. And if green plants had perished, it would inevitably have shared their fate. This is a single, indivisible, natural phenomenon.

The animal realm does not manifest life in itself. An animal organism disperses the energy accumulated by green chlorophyll-containing organisms inside its physiological machine. But the totality of animals, especially civilized humanity, corresponds to the same energy manifestations that are so characteristic of green plants. In general, all the animals and plants – all living nature – demonstrate a natural phenomenon that contradicts the usual formulation of Carnot's principle in its effect. As the result of life and all its manifestations, usually *an increase of active energy takes place in the Earth's crust.*

Paying attention to the whole biogeochemical work produced by living organisms – inseparable from them and created by them at the expense of the energy they capture – we see that a single complex totality of self-dependent organisms appears in this way. The active energy of living organisms increases while the same initial, continuous, and steady solar energy increases. It increases over the course of geological time. This increase of active energy is expressed, for instance, in the increase of consciousness, and in the growth of life's influence in the biosphere in geochemical processes. An example of such a geological power that shows it clearly and that has proceeded slowly in geological time is the civilized humankind characteristic of our psychozoic era.

But the same is shown to us by the evolutionary process of species, which is inseparably connected with the growth of active geological energy and with the complete change of the biosphere. The effects of life upon the biosphere increase with the uniform flow of active (solar) energy. Living matter creates and accumulates it, and does not disperse it. The same is shown by the increase of the geochemical functions of life in the course of geological time, by the greater variety of its morphological forms, which must inevitably be related to the increase of chemical variety.

Due to the existence of life, the entropy of the universe should decrease in biospheric phenomena, and not increase. This empirical generalization brought about new speculations. The German physicist, F. Auerbach, regarded it as a new principle contradicting that of entropy. He called it ectropy. He and other scientists tried to draw cosmogonic conclusions from it. However, there is nothing to make us put forward new hypotheses. Clausius' entropy does not really exist; it is not a fact of being, but a mathematical expression, useful and necessary when it allows expression of natural phenomena in mathematical language. It is correct only as far as its premises are concerned. The deviation from Carnot's principle by such an essential phenomenon as living matter and its influence upon the biosphere shows that life does not stay within the premises for which entropy is stated.

The essential fact is that life in its most distinct manifestations is inseparably connected (like radioactivity) with the microscopic view of the world, where such regulations as those of thermodynamics do not exist. It is in the biosphere that this close connection of life with phenomena taking place outside the usual gravitational field must manifest most distinctly, as the connection of life with the biosphere is inseparable in the phenomena studied by us, and manifests itself in the great migrations of Earth's matter. The special capacity of the biogeochemical phenomena shows that the phenomena of the atomic world – of the microscopic view of the world – can play a leading role in discovering the final results of life in the biosphere.

At present this is the direction in which physical thought is working, and the possibility of this very explanation shows the energetic peculiarity of life phenomena in the biosphere – its deviation from Carnot's principle. As we shall see in the next essay, in the twentieth century new natural phenomena were discovered – those of radioactivity – inducing the same contradictions to Carnot's principle as it is usually understood, and that are related to the same manifestation of the microscopic view of the world in the macroscopic phenomena of the Earth's surface.

21   the free energy of the biosphere and living matter

But let us return to the geochemical history of carbon and living matter, a very clear expression of the same energetic manifestation of life. Living matter increases the active energy of the Earth's crust in two different ways due to its reversible processes. First, by releasing gases (their close connection with life has already been mentioned), and as a consequence creating the terrestrial atmosphere, as its dominant gases such as nitrogen, oxygen, and carbon dioxide are biogenic.

As for the creation of the planet's free, active energy, the transformation of solar radiant energy into free oxygen through living matter is most important. Free oxygen impacts the whole planet's surface and gives it quite special qualities that can't be observed anywhere else. Another phenomenon of the same character leading to the same result – the growth of the planet's active energy – is the dispersion of life over all the planet by means of proliferation.

The organism's proliferating apparatus is a specific mechanism for the dispersion of the geochemical energy of life, regulating elements' migration in the biosphere and thus in the entire Earth's crust. The speed of dispersion in organisms that are the most adapted for this purpose reaches their physical limit. Thus, living matter becomes the regulator of the biosphere's active energy. It distributes the latter more or less uniformly over the Earth's surface. Therefore, the surface layer of the planet becomes, through living matter, a field of manifestation of kinetic and chemical energy.

This dispersion of free energy is apparently continuous and unchangeable, like solar radiation. At the same time, due to reversible chemical processes, living matter gathers the radiant energy of the Sun and passes it to the deeper layers of the Earth's surface as chemical compounds, which under certain conditions can release chemical energy, mostly in the form of organogenic carbonaceous bodies and carbonates. The quantity of this accumulated potential energy increases over the course of time to limits determined by geological conditions. In the same way, the radiant energy of the Sun is concentrated through living matter in all chemical compounds formed by free oxygen (always biogenic), and apparently in all vadose minerals, since they are always more or less closely connected with life.

This energy is gradually separated in the course of the decomposition of compounds, while they turn into stable compounds under new thermodynamic conditions. For carbon, we know two forms of stable compounds that contain almost no accumulated energy of solar rays. These are pure carbonic acid and graphite (the native carbon). Pure gaseous carbonic acid [CO_2] returns to the surface and re-enters the cycle of life, whereby the cyclical process is completed. Graphite remains absolutely inert in the biosphere only if it does not oxidize in a dispersed state under the influence of bacteria. Having left the biosphere and reached the juvenile areas or magmas, it may again enter into different chemical reactions under the influence of the free energy of these areas' high temperature.

Those chemical reactions can take place for stable carbonaceous matter only due to its exceptional dispersion. The dispersion of carbon is apparently related to living matter; that is, it is an event produced by solar energy. In the total economy of the Earth's crust, large accumulations of limestone, coal, or crude oil are the dominating form of carbon. But their small masses dispersed as "traces" that are found everywhere, play a much more important role. They are chemically active, they oxidize, and they become centers of chemical activity. Due to their small size, due to the molecular power of their surfaces, they bring about chemical reactions that are impossible from the ordinary energetic point of view, and thereby manifest themselves in truly grand geochemical phenomena.

From the geological standpoint, it is necessary to indicate that this state of dispersion is concentrated neither in the ocean nor in the seas where the greatest mass of living matter is accumulated, but on continents and islands, and especially in the gaseous atmosphere. Here we see a new form of the geochemical role of these parts of the Earth's crust.

22 geochemical activities of man

The geochemical cycle of carbon, as well as the geochemical history of other chemical elements, does not remain unchanged throughout geological time. During the evolution of plant and animal species, the chemical molecules forming them do not remain the same. But this change of the chemical composition has manifested itself in the course of geological time only inside living matter. In the inert matter outside it, the same associations of the same minerals were being created from the Archaic era to the Pleistocene.

But in our geologic era, in the psychozoic era – the era of reason – a new geochemical factor of paramount importance appears. During the last ten or twenty thousand years, the geochemical influence of mankind, which has captured green living matter by means of agriculture, has become unusually intense and diverse. We see a surprising speed in the growth of mankind's geochemical work. We see a more and more pronounced influence of consciousness and collective human reason upon geochemical processes. Man has introduced into the planet's structure a new form of effect upon the exchange of atoms between living matter and inert matter. Formerly, organisms affected the history only of those atoms that were necessary for their respiration, nutrition, and proliferation. Man has widened this circle, exerting influence upon elements necessary for technology and for the creation of civilized forms of life. Man acts here not as *Homo sapiens*, but as *Homo sapiens faber* [faber = fabricator, maker].

He expands his influence to all elements. He changes the geochemical history of all metals, he makes new compounds, and he reproduces them in quantities of the same order as mineral products of natural reactions. This is a fact of exceptional importance in the history of chemical elements. For the first time in our planet's history we see the creation of new chemicals – an incredible change of the Earth's face. From the geochemical standpoint, all these products – the masses of free metals such as metallic aluminum, which had never existed on Earth, or those of iron, tin and zinc, the masses of carbonic acid produced by calcinating lime or by burning coal, the great quantities of sulfuric anhydride or hydrogen sulfide formed in the course of chemical and metallurgical processes, and the increasing quantities of other technical products – do not differ from minerals. They change the eternal motion of geochemical cycles. With the further development of civilization, the influence of these processes must increase, atomic migration on a biogenic basis will expand, and at the same time the number of atoms used by it will grow.

Evidently, this is not an accidental fact; it was predestined by the whole paleontological evolution. It is a natural fact like all others. But we see a new phenomenon here in which living matter seems to act in sharp contradiction to Carnot's principle. Where will this new geological process stop? And will it stop at all? Poets and philosophers give us answers that often do not seem incredible and impossible to a man of science. The study of geochemistry proves the importance of this process and its deepest connection with the entire chemical composition of the Earth's crust. It is still in a state of evolution, the final result of which is hidden from us so far. But as it is, it is a factor changing the reversible geochemical cycles of all elements more and more sharply. It introduces new compounds into the Earth's crust, and these compounds are still less stable in the thermodynamic conditions of the Earth's crust than those existing before; they are a source of a more intense active energy, increasing the free energy of the Earth's crust, which has remained unchanged from time immemorial.

Man always increases the number of atoms leaving the ancient cycles – the geochemical "eternal" cycles. He intensifies the breach of these processes, introduces new ones, and interferes with old ones. With Man, an enormous geological power has appeared on the surface of our planet. The balance of the migrations of elements that had been established in the course of geological time is being broken by the reason and activities of Man. At present we are changing the thermo-dynamic equilibrium inside the biosphere in this way.

the biosphere

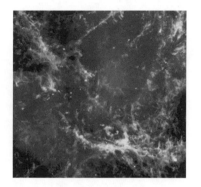

the biosphere in the cosmos

1 the biosphere in its cosmic environment

The face of the Earth, its image in the cosmos as seen from outside, from the depths of infinite celestial space, seems to us absolutely unique, inimitable and distinct from all other heavenly bodies.

The face of the Earth exhibits the surface of our planet, its biosphere – its outer domain separating it from its cosmic surroundings. The face of the Earth becomes visible due to light penetrating into it from celestial bodies, especially from the Sun. From all space it gathers an infinite diversity of radiation of which the luminous rays visible to us are only a small part.

Only a few kinds of invisible radiation are known to us at present. We have just begun to realize the diversity, to understand how fragmentary and incomplete our notions are about the realm of radiation that surrounds us and permeates us in the biosphere, and about the fundamental role of radiation in the processes around us – something that is hardly understandable by an intellect accustomed to a different picture of the world.

Radiations of immaterial substance [i.e., pure energy] embrace not only the biosphere, but all conceivable space. Around us, inside us, everywhere, radiations are diffused, incessant, ever changing, coinciding and crossing, with wavelengths ranging from a millionth part of a millimeter up to several kilometers.

The whole of space is full of them. It is difficult, perhaps even impossible, for us to imagine these surroundings, the cosmic environment of the world in which we live and in which, at one and the same place and at one and the same time, we distinguish and measure ever new radiation in the course of improving our methods of research. Eternally changing and continually filling space, they create a sharp distinction between the cosmic environment devoid of matter and an ideal geometric space.

These radiations are of different types. They reveal changes of material bodies and their surroundings. We perceive some of them as energy; that is, transmission of state. But besides these, in the same cosmic space, another type of radiation exists, the velocity of which is often of the same order – the radiation consisting of extremely small, quickly moving particles such as electrons.

These are two sides of the same phenomenon, and they penetrate into each other. The transmission of state is a manifestation of the movement of single entities, be it quanta [now called *photons*], electrons, magnetons, or single entities or charges.[1] Separate elements are connected with these entities, although the latter may themselves remain immobile.[2]

---

1. Perhaps Vernadsky means that when an atom or molecule changes its quantum state, it gives off radiation, which can be pure radiant energy with a wave and particle (photon: no rest mass) character, or it can be a different, usually much heavier particle such as an electron, which carries a negative charge. –*Ed.*

2. Vernadsky may be thinking of particles given off in radioactive decay, newly discovered shortly before his time: beta rays, which are electrons; or alpha particles, which are equivalent to the nucleus of a helium atom. These particles with considerable rest mass do not travel at the speed of light. –*Ed.*

Particle radiation is a manifestation of transmission of separate elements of these entities. These particles [i.e., "matter radiation"], as well as radiation performing the transmission of states [i.e., "electromagnetic radiation"], may pass through material bodies constituting the world. They may transform the phenomena in the surroundings they appear in as radically as the forms of energy do.

2   At present our knowledge on this subject is far from being satisfactory. We can as yet ignore the particle radiations while considering the geochemical phenomena of the biosphere, but in our calculations we must always take into account the radiations of state transmission; that is, the forms of energy.

According to their particular forms – for instance, their wavelength – they appear to us as light, heat [infrared], or electricity, and they can transform the material surroundings – our planet and the bodies constituting it – in different ways. From the standpoint of wavelength there is an immense range of these radiations. At present it embraces about 40 octaves [an octave is a doubling of frequency]. We can get an idea of this number by remembering that the visible part of the Sun's spectrum represents one octave.

Obviously our knowledge does not cover the whole range of octaves. In the course of scientific creative work the field of radiation becomes wider and wider . . . But our scientific notions of the cosmos, our habitual picture of the world, include only few of those forty octaves, the existence of which is without doubt.

The cosmic radiations received by our planet that are, as we shall see, creating its biosphere, extend over only 4.5 of the forty known octaves. It seems improbable that the other octaves should be missing in cosmic space. We regard their absence as an illusion caused by the absorption of the rays by the rarefied material of the upper layers of the atmosphere.

In the most familiar cosmic radiations – those from the Sun – we distinguish one octave of light, three octaves of heat [infrared] and half an octave of ultraviolet radiation – the last one being a small remnant that passed through the stratosphere.

3   Radiation from the cosmos throws an eternally and continually powerful stream of energy onto the face of the Earth, conferring quite a peculiar and novel character on the parts of the planet bordering cosmic space. Due to cosmic radiation, the biosphere acquires characteristics that are novel, unusual, and unfamiliar to the Earth's matter – the face of the Earth reflected in its cosmic environment shows a new picture of the Earth's surface as it is transformed by cosmic forces.

As a result of these radiations the substance of the biosphere is penetrated by energy – it becomes active, it gathers and distributes the energy received in the form of radiation and eventually turns it in terrestrial organisms into free energy capable of performing work.[3] The external envelope of the Earth formed from this living substance can therefore not be regarded as a domain of sheer substance – it is the domain of energy, the site of transformation of the planet by external cosmic forces.

---

3. This is the thermodynamic definition of "free energy": energy capable of doing "work." (It is not just energy that doesn't cost anything!) –*Ed.*

The face of the Earth is transformed by these forces and to a great extent molded by them. Not only is it a reflection of our planet, a feature characteristic of its substance and energy, but at the same time it is a creation of the external forces of the cosmos. The history of the biosphere is therefore sharply distinguished from that of the rest of the planet, and the role it plays in the planetary mechanism is quite exceptional. It is as much, or even more, the creation of the Sun as it is a manifestation of terrestrial processes.

Great ancient religious intuitions of humanity that regarded terrestrial creatures, especially human beings, as "children of the Sun" were much nearer to the truth than theories considering them to be merely an ephemeral product of blind and accidental transformations of the Earth's matter and forces. Terrestrial creatures are the products of a complicated cosmic process – they constitute a regular and necessary part of a harmonious cosmic mechanism in which, as we know, there is no place for chance.

4   We are led to the same conclusion by our concepts of the matter constituting the biosphere, which concepts have changed greatly in recent years. They make it inevitable for us to consider the substance of the biosphere as a manifestation of cosmic mechanisms. This is in no way due to the fact that part of the material of the biosphere, perhaps the larger part, is of non-terrestrial origin and comes from outside, from cosmic space. For this substance from outside – cosmic dust and meteorites – is of the same structure as the terrestrial substance. Its unexpected structure is just being revealed to us, and we have not come to a definite and complete understanding of it yet. Still, the modifications of our views are so profound and change our understanding of geological phenomena to such an extent, that on entering this field of terrestrial phenomena we must begin by taking them into consideration.

The similarity between the structures of cosmic substance reaching us and the structure of terrestrial substance is not limited to the biosphere, the thin outer film of the planet. It is the same for the whole crust of the Earth – for the kilometers of lithosphere of which the upper part is the biosphere. Lithosphere and biosphere gradually and inseparably merge with each other.

Undoubtedly, the substance of the deeper parts of the planet has the same character, though it is different in its chemical composition and presumably always foreign to the Earth's crust. Therefore it may be ignored while studying the phenomena observed in the biosphere. The substance of the terrestrial domains that are lower than the crust of the Earth hardly penetrates it in large amounts during short periods of time.

5   For a long time there has been no doubt that the chemical composition of the Earth's crust is determined purely by geological factors, and is the product of interactions of various geological phenomena at different scales. The explanation of this was sought in the joint action of the same geological phenomena we are observing at present in our surroundings – in the chemical and solvent action of water, atmosphere, organisms, volcanic eruptions etc. The Earth's crust seemed to have acquired its present chemical composition, both quantitative and qualitative, due to the interaction of the same geological processes and of the same chemical elements throughout the course of geological time.

Such an explanation presented many difficulties and was not the only one. According to more complex ideas, geological phenomena change throughout time and thereby account for the present chemical composition. For instance, this composition was regarded as the residue from ancient periods in the Earth's history that were unlike the present day – the crust of the Earth was regarded as a kind of slag formed from the formerly molten mass of our planet in complete accordance with the laws governing the distribution of chemical elements in such molten masses as they solidify under falling temperature.

In order to explain the predominance of the lighter elements in the crust, one had to refer to still more ancient periods in the Earth's history preceding the formation of the crust – the space periods. It was thought that at the time when the molten mass of the Earth had been appearing out of a nebula, the heavier elements concentrated around the center. In all these notions the composition of the Earth's crust was connected with geological phenomena. The elements took part in them according to their chemical properties when they could form chemical compounds, and according to their atomic mass under high temperature when all compounds seemed unstable.

6 The laws governing the chemical composition of the Earth's crust that have been revealed now are in flagrant opposition to these explanations. At the same time, the general picture of the chemical composition of all other celestial bodies shows us a complexity, uniqueness, and regularity we could not even suspect before.

In the structure of our planet, and especially of its crust, there are indications of phenomena that far transcend its limits.

We are unable to understand them without getting away from purely terrestrial, even planetary phenomena, and turning to the structure of cosmic matter as a whole – to its atoms and to their modifications in cosmic processes. Many different indications that are hardly embraced by theoretical thought are quickly accumulating in this field. Their significance is only beginning to be realized. Not always can they be formulated in a clear and definite way, and usually no conclusions are drawn from them. Nevertheless, the great significance of these phenomena should not be forgotten. It is necessary to take into account the unexpected consequences of these new facts.

Three fields of phenomena can be distinguished at present:

1. the special status of the elements of the terrestrial crust in the Periodic Table,
2. their complexity, and
3. the irregularity of their distribution.

For instance, elements with even atomic numbers definitely prevail in the mass of the Earth's crust. No geological reasons known to us can account for that fact. Besides, it was immediately discovered that the same phenomenon was manifesting even more distinctly in meteorites, the only extraterrestrial cosmic bodies accessible to direct scientific research.

Other facts are perhaps even more mysterious. Any attempt to explain them by geological reasons contradicts the phenomena known in this field. We cannot understand the immutable complexity of the terrestrial chemical elements and the certain constant correlations in the number of their isotopes. The study of isotopes of the chemical elements in meteorites showed the similarity of the compounds in these bodies, although their history and location in space are obviously different from Earth's.

It has also become evident that it is impossible to explain the definite composition of the Earth's crust and the whole planet by the different atomic masses of the elements comprising it. Not geological but some other causes are to account for the difference in composition between the Earth's crust and the Earth's core – the discovered similarity between the composition of meteorites and that of the deeper layers of our planet cannot be accidental. The reason for the prevalence in the Earth's crust of relatively light elements, but rather heavy iron as well, should not be sought in geological or geochemical phenomena – not only in terrestrial history. It lies deeper; it is related to cosmic history and maybe to the structure of chemical elements.

New unexpected proof of this conclusion can presently be found in the discovered similarity between the outer parts of the Earth (i.e., the Earth's crust), the Sun, and the stars. As early as 1914, Russell pointed out the similarity between the composition of the Earth's crust and the Sun (i.e., its outer layers, which we are studying). These interrelations are shown even more distinctly in the new investigations of the spectra of stars. They show the following decreasing order of the prevalence of chemical elements: Si – Na – Mg – Al – C – Ca – Fe (>1%); Zn – Ti – Mn – Cr – K (0.1%–1%). We see a clear analogy with the same order of chemical elements in the Earth's crust: O – Si – Al – Fe – Ca – Na – K – Mg.

These works present the first achievements in this vast and novel field of phenomena. They still require proof and checking, but at present we cannot ignore them or fail to take into account the fact that the first obtained results show a striking similarity in composition between the surfaces of celestial bodies: the Earth, the Sun and the stars. The outer parts of celestial bodies are related directly to cosmic space. By way of radiation they exert a reciprocal effect on one another. The explanation of this phenomenon may be found in the interchange of matter that is likely to take place between these bodies in the cosmos. The deeper parts of cosmic bodies probably present a different picture. Meteorites and the Earth's inner masses are entirely different in composition from the outer layers known to us.

7 Thus our ideas about the nature of our planet's composition, especially that of its crust and its outer envelope (the biosphere), undergo sudden changes. We begin to see them not as purely terrestrial phenomena, but related to the structure of the atoms, to their location in the cosmos, and to their evolution in the history of the cosmos. Even though we are unable to understand these phenomena, we have taken the right direction and entered a new field of phenomena that is quite different from the one we were so long trying to attach to the chemistry of the Earth. Now we know where to search for the solution of the problem facing us, and where it is hopeless to search for it. So our understanding of the observed phenomena undergoes radical changes.

The upper surface film of our planet – the biosphere – is the place for us to search not only for the reflections of accidental unique geological facts, but also for manifestations of the structure of the cosmos related to the structure and history of chemical atoms. The phenomena in the biosphere cannot lead to understanding of the biosphere unless one takes into account the obvious bond that unites it with the structure of the entire cosmic mechanism. We can trace this bond through the innumerable facts of its history that are familiar to us.

8   the biosphere as the reason for the transformation of cosmic energy

The biosphere may, by reason of its essence, be regarded as a part of the Earth's crust occupied by transformers that turn the radiations from the cosmos into active terrestrial energy: electrical, chemical, mechanical, thermal energy, etc. Radiation from all the celestial bodies encompass the whole biosphere and penetrate it and everything in it. We perceive and realize the existence of only a small part of this radiation, almost exclusively that from the Sun. But we know about the existence of other types of radiation originating in the remotest parts of the cosmos. For instance, the stars and the nebulae are incessantly sending light radiation to our planet.

Everything speaks in favor of the fact that the radiation penetrating the upper layers of the atmosphere discovered by V. Hess originate beyond the limits of the Solar System. Their origin is looked for in the Milky Way, in the nebulae, in the stars of the Mira Ceti type. Maybe it is the Milky Way that emanates the mysteriously penetrating radiation, which is so bright in the upper layers of our atmosphere.

In the future, radiation will surely be taken into account and explained, but undoubtedly it is not such radiation but the rays of the Sun that determine the main characteristic features of the mechanism of the biosphere. The influence of solar radiation on terrestrial processes is studied well enough to achieve a first, but deep and precise notion of the biosphere as a terrestrial and cosmic mechanism. The Sun has completely transformed the face of the Earth – it has permeated and encompassed the biosphere. To a large extent, the biosphere is a manifestation of its radiations – it is a planetary mechanism that converts these radiations into new diverse forms of terrestrial free energy and thereby changes the history and destiny of our planet completely.

We are already aware of the great significance in the biosphere of the short ultraviolet waves of solar radiation, the longer red radiant-heat waves and the intermediate rays of the visible light spectrum. We can detect the elements that fulfill the role of trans-formers of each of these different systems of solar oscillations.

It is slowly and with difficulty that we begin to realize the mecha-nism of this transformation of solar energy into terrestrial forces within the biosphere. We are accustomed to seeing its phenomena from another standpoint – it is veiled by the infinite variety of color, form, and movement inherent in nature. We ourselves form an integral part of it through our lives. Centuries and millennia had passed until human thought was able to grasp the signs of a single mechanism in the seemingly chaotic picture of nature.

9   The transformation of the three systems of solar radiation into terrestrial energy takes place partially in the same regions of the biosphere, but sometimes certain fields are distinguished in which some particular kind of transformation prevails. The apparatus for the transformations always consists of natural bodies and is quite different for ultraviolet, luminous, and thermal rays. Certain short ultraviolet rays are completely or partially absorbed by the upper rarefied layers of the Earth's gaseous envelope – the stratosphere – and possibly by the "free atmosphere" that is still higher and poorer in atoms.

This "absorption," or "detainment," is related to the transformation of the radiant energy of the short waves. Changes of electromagnetic fields, decomposition of molecules, various phenomena of ionization, and formation of gas molecules of new chemical compounds occur in these regions under the influence of ultraviolet rays. The radiant energy is partly transformed into various kinds of electric and magnetic phenomena, and partly into related molecular, atomic and peculiar chemical processes of rarefied gaseous states of substance.

We see these regions and these bodies as Aurora Borealis, summer lightning, zodiacal light, a glimmering of the sky which becomes visible only on dark nights but still constitutes a significant part of the nocturnal sky's illumination – we see them as luminous clouds and other diverse reflections of the stratosphere and outer limits of the planet in the picture of our terrestrial world. Our instruments disclose this mysterious world of phenomena as electric, magnetic, radioactive, chemical and spectroscopic manifestations, in its incessant movement and variety, which is beyond our imagination.

These phenomena are not a consequence of the transformation of the Earth's surroundings by the solar ultraviolet rays alone. We must take into account the whole complicated process. All forms of solar radiant energy beyond the limits of 4.5 octaves, which penetrate into the biosphere (§2), are "detained" here; i.e., transformed into new, terrestrial phenomena. The powerful flows of electrons that are incessantly emanated by the Sun, as well as the material particles – cosmic dust and gaseous bodies – are captured by Earth's gravity and supply the Earth with new sources of energy.

The significance of these phenomena for the history of our planet is gradually penetrating into common thinking. Short luminous waves of 180–200 nanometers destroy all living organisms – longer and shorter waves are harmless to them.[4] By absorbing the short waves completely, the stratosphere protects the lower layers of the Earth – the sphere of life. It is very significant that the absorption of these rays is related to ozone (the ozone screen), the formation of which depends on free oxygen, which itself is a product of life.

10   Although we are just beginning to realize the importance of the transformation by ultraviolet rays, the role played by solar heat, mainly in the form of infrared rays, has long been recognized. It turns out to be exceedingly important for studying the influence of the Sun on geological, and even on geochemical processes. The importance of radiant solar heat for the existence of life is obvious and without doubt. The transformation of the thermal radiant

---

4. Vernadsky's original text said: "180–200 megaherz," or a frequency of 180–200 x 10^6 s^{-1}. That is in the range of broadcast radio waves, which are utterly harmless. (The frequency of visible light is in the range of 4.0 to 7.0 x 10^{14} s^{-1}.) If Vernadsky really meant to describe the radiation in units of wavelength rather than frequency, 180–200 nanometers would be in the very short ultraviolet part of the spectrum – and very deadly. Thus, it seems apparent that Vernadsky meant wavelength in nanometers rather than frequency in megahertz. –*Ed.*

energy of the Sun into mechanical, molecular (condensing, etc.) and chemical energies is also incontestable.

Such transformations can be seen everywhere and their exisence requires no explanation. We observe them in the life of organisms, in the movement and work of winds or ocean currents, in waves and surf, in the destruction of rocks and action of the glaciers, in the movement and formation of rivers, and in the colossal work of snow and rain. We are usually less aware of the role played by the liquid and gaseous parts of the biosphere in accumulating and distributing heat and thus transforming the radiant thermal energy of the Sun.

Our atmosphere, ocean, rivers and lakes, and rain and snow act as transformers. The ocean, thanks to the unique, exceptional thermal properties of its water and to the character of its molecules, plays the immensely important role of heat regulator and hereby acts upon the endless phenomena of weather and climate and their associated processes of life and weathering. Heating up quickly by reason of its large thermal heat capacity, the ocean slowly gives away its heat because of the character of its heat-conductivity. It transforms the absorbed radiant heat into molecular energy through evaporation, into chemical energy through living matter with which it is permeated, and into mechanical energy through ocean currents and breakers. The thermal functions of rivers, precipitation, and aerial masses with their heating and cooling, are of the same direction and of comparable scale.

11 Ultraviolet and infrared solar rays only have an indirect effect on the chemical processes of the biosphere. They are not the main source of its energy. The chemical energy of the biosphere in its active form is received from the radiant energy of the Sun by the aggregation of living organisms on Earth – its living matter. With the help of photosynthesis, solar rays create an infinite number of new chemical compounds, many millions of different combinations of atoms; it covers the Earth with a powerful layer of molecular systems that very easily produce new compounds rich in free energy in the thermodynamic field of the biosphere. These are unstable in this field and inevitably pass on to new forms of stable equilibrium.

This process takes place with an amazing quickness, and this type of transformer constitutes a mechanism entirely different from the terrestrial bodies transforming the short and long waves of solar radiation into new forms of energy. We explain the transformation of ultraviolet waves by their effects upon matter – those atomic systems formed independently of them. As for the transformation of thermal radiations, we relate it to molecular structures created beyond their direct influence. But photosynthesis as it is observed in the biosphere is related to special, exceedingly complicated mechanisms, which it creates by itself on condition of the simultaneous manifestation and transformation of ultraviolet and infrared solar radiation in the environment.

The mechanisms for transforming energy created in this way – living organisms – are very special formations, entirely distinct from all atomic, ionic, or molecular systems that build up the substance of the Earth's crust outside the biosphere, and partially the substance of the biosphere itself.

The structures of living organisms are of the same type as those forming inert matter, but more complicated. Because of the changes they create in the chemical processes of the biosphere, they cannot be looked upon as mere inert components of these structures. Their energy character as manifested in their propagation is from a geochemical standpoint incomparable to the inert structures constituting both dead and living matter.

The mechanism of chemical action of living matter is unknown to us, but it is obviously becoming clear that from the standpoint of energy phenomena in living matter, photosynthesis takes place not only in special chemical surroundings, but also in a special thermodynamic field that is different from the thermodynamic field of the biosphere. After the organism's death, the compounds that are stable in the thermodynamic field of living matter enter the thermodynamic field of the biosphere where they prove unstable and serve as sources of free energy. The field of phenomena inside an organism (the thermodynamic field of living matter) is thermodynamically and chemically different from that of the biosphere.

12   an empirical generalization and a hypothesis

Obviously, such an understanding of energy phenomena as far as it manifests itself in geochemical processes expresses the observed facts correctly. But we cannot assert it because here we come across a peculiar state of our knowledge in the sphere of biological sciences as compared to the sciences dealing with inert matter.

We have already seen that in the latter it has been necessary to reject previous ideas of the biosphere and the structure of the Earth's crust – to put aside the explanations of purely geological character that reigned supreme and seemed certain for many generations (§6).

Explanations that had seemed scientifically and logically inevitable eventually proved to be an illusion, and the phenomenon has
acquired an entirely unexpected form. In the sphere of studying
life, the situation is still more difficult, because there is hardly any
branch of natural sciences with fundamental principles that are so
full of philosophical and religious concepts alien to science.

Our ideas of living matter are influenced considerably by philosophical and religious studies and achievements. For centuries, even the
statements of precise naturalists had an imprint of these intuitions
of the human mind about the cosmos – intuitions alien to science in
essence, although deep and precious. It became difficult to retain a
scientific approach while studying this field of phenomena.

13   The two main views on life, both the vitalistic and the mechanistic
ones, are results of these philosophical and religious ideas and not
of scientific deduction. They both hinder the study of life phenomena and confuse empirical generalizations.

The first tries to explain the phenomena of life in a way that remains
outside the world of models that represent our scientific generalizations concerning the cosmos. Due to the character of these notions,
they are devoid of any creative significance in a scientific sense;
they are scientifically fruitless.

No less dangerous are the ideas of mechanical character, seeing
in living organisms nothing but a game of physical and chemical
powers. They limit the sphere of scientific search and predetermine
its results. They bring an element of guessing into the scientific
sphere, which obscures scientific understanding. Of course, if the

guess had been successful, scientific processing would soon have smoothed all its roughness. But the guess turned out to be too closely connected with abstract philosophical constructions foreign to the reality studied by science, and this leads to simplified ideas of life that prevent us from understanding the complexity of phenomena. Up until now – for centuries – this view has not brought any progress to the understanding of life.

That is why it should be correct to put aside both these ways of explaining life, and to approach its phenomena from a merely empirical standpoint, taking into account the impossibility of giving it an "explanation;" that is, to find a place for it in our abstract cosmos made up of models (hypotheses). This approach is becoming more and more dominant in scientific research.

Now only an empirical approach to the phenomena of life is sure to bring success, provided the hypotheses are not taken into consideration.[5] This is the only approach that can find new features in life phenomena, which will either enlarge the sphere of physical and chemical forces familiar to us until now, or introduce a new principle or axiom into science – a new notion that cannot be proven or regarded as a mere consequence of the familiar axioms, and that will stand together with those comprising our scientific world of matter and energy. Then it will be possible to bring in hypotheses

---

5. Considering the paragraphs above this one, Vernadsky here must be referring specifically to the hypotheses of vitalism and mechanism. Yet in other places (e.g., §15), he would convince us that hypotheses ("mere guess-work") in general are to be avoided in scientific work. This is a strange idea to contemporary scientists, who pursue their fields by making and testing hypotheses. Yet he ends this paragraph with a positive use of "hypotheses": ". . . bring in hypotheses to connect these phenomena. . .." Hence, Vernadsky seems to be saying here that we should avoid the hypotheses of vitalism and mechanism if we want to make progress. –Ed.

and to connect these phenomena to our models of the cosmos, like the discovery of radioactivity is connected to the reality of atoms.

14   A living organism of the biosphere must now be empirically studied as a special body that cannot be completely reduced to physical and chemical systems we know. Science at present is unable to decide whether in the future it will be possible to reduce an organism in this way. It seems possible, but in our empirical study of natural phenomena we must not forget another possibility – the fact that this very task which has been posed by many scientists, may turn out to be as illusory as the problem of squaring the circle. In the sphere of biology we have come across such problems more than once.

The geological sciences require an empirical standpoint without mechanistic and vitalistic notions even more than the biological ones.

In one of them, namely in geochemistry, one has to face the phenomena of life very often. Here, organisms in the form of their aggrate – living matter – are one of the principal working factors.

This living matter gives the biosphere an absolutely unusual and unique image in the world. Whether we want to or not, we cannot but distinguish two kinds of matter in it – the living and the inert matter, which influence each other, but which at some points in their geological history are separated by an unbridgeable gulf. No doubts can ever arise about these two types of biospheric matter belonging to different categories of phenomena, which cannot be united.

Their main difference, no matter what it is, presents not only an empirical fact, but also one of the most important empirical generalizations of the natural sciences. The significance of this generalization and the significance of all generalizations in science are often forgotten. Instead, under the influence of routine and philosophical constructions, these empirical generalizations are identified with scientific hypotheses of vitalism and mechanism. Dealing with the phenomena of life, it is especially necessary to avoid this deeply rooted bad habit.

15 A great difference exists between empirical generalizations and scientific hypotheses, and the exactitude of their conclusions is far from being the same.

In both cases, for empirical generalizations as well as for scientific hypotheses, we use deduction for drawing conclusions that can be checked by studying real phenomena. In a historical science like geology, this checkup is performed by means of scientific observation.

But the difference is that an empirical generalization is based on facts gathered by inductive methods, without going beyond their limits and without caring about whether the received conclusions correspond or do not correspond with other existing notions about nature. In this respect, an empirical generalization does not differ from a scientifically determined fact; their coincidence with our scientific notions about nature does not interest us, and their discrepancy with these notions means a scientific discovery. In an empirical generalization, although certain features of the phenomenon are considered principal, all the others also matter. Establishing a scientific fact by empirical generalization means that the conclusion is always drawn on the basis of the entire phenomenon.

An empirical generalization may exist for a very long time without yielding any hypothetical explanations, remain uncomprehended, and still exercise a great positive influence upon the understanding of natural phenomena. But then a moment comes when it begins to be seen in some new light, becomes part of the sphere of hypothesis – changes our schemes of the world, and is changed itself. Then it very often turns out that an empirical generalization is not what we have thought of it, or actually much more than we have thought of it. A typical example of such a story of empirical generalization is one of the greatest: the Periodic Table of D. I. Mendeleyev, which after the discovery of J. Moseley made in 1915, created a vast sphere of scientific hypotheses.

16   A hypothesis or a scientific construction is built up in a quite different way. For a hypothesis, one or several significant features of a phenomenon are taken into account, and only on their basis is the idea about the phenomenon created – without paying attention to its other aspects. A scientific hypothesis always goes beyond the limits of mere facts which have served as a foundation for its building, and that is why, in order to be strong enough, it must inevitably be related to possibly all the dominating theoretical constructions concerning nature, and not contradict them.

17 Thus an empirical generalization being a precise consequence of facts does not require checking. Only empirical generalizations based on series of well-known facts, neither hypotheses nor theories, comprise the foundation of my further narration.

I introduce the following statements:

1. During all geological periods, there have not been and there are not now any traces of abiogenesis (i.e., direct creation of living matter from dead, inert matter).
2. During the whole length of geological time, no azoic (i.e., lifeless) geological eras have been observed.
3. This results in the fact that, on the one hand, the contemporary living matter is genetically related to living matter of all geological eras, and, on the other hand, during all this time the conditions of the terrestrial environment have been favorable for its existence; that is, they have always been close to the contemporary reality.
4. In the course of all geological time, there have been no sharp changes of any kind in the chemical influence of living matter upon its surroundings; during all this time the same processes of weathering took place on the Earth's surface, i.e., the average chemical composition of living matter and the Earth's crust was similar to that observed nowadays.
5. The constancy of the weathering processes results in the unchangeable quantity of atoms seized by life; that is, no great changes have occurred in the quantity of living matter.[6]

---

6. There have only been some insignificant oscillations around a constant average number.

6. No matter of what the life phenomena consist, the energy produced by organisms is mainly and perhaps wholly the radiant energy of the Sun. Through organisms the Sun regulates the chemical manifestations of the Earth's crust.

18 These empirical generalizations being accepted as the basis of our judgements inevitably result in the statement that a series of problems posed by science, especially in its philosophical adaptations, escapes the sphere of our attention, because these problems are not consequent to the empirical generalizations and cannot be built up without hypothetical assumptions. For instance, no consideration may be given to the problem of the beginning of life on Earth, if there was a beginning to all cosmogonic ideas of the past lifeless state of the Earth and to the possibility of existence of abiogenesis during the hypothetical cosmic periods of the Earth's history.

These issues – abiogenesis, the beginning of life and the existence of lifeless periods in the history of the Earth's crust – are so closely connected with the dominant scientific and philosophical constructions and deeply permeated by cosmogonic hypotheses, that they may seem logically inevitable to many people. But studying the history of science shows that these issues have penetrated into science from outside, were born outside science in religious or philosophical searches of humanity. And this can be clearly seen while comparing them with the empirical sphere of precise facts, which builds up science.

All the precise facts known to us will never change, even if all those problems are solved in a negative way; that is, if we admit that life has existed forever and had no beginning, that life – a living organism – has never and nowhere originated from dead matter, and that in the geological history of the Earth there have been no periods devoid of life. But we shall have to build new cosmogonic theories to replace those dominant now and to apply another kind of mathematical or scientific processing to some philosophical and religious constructions left aside by scientific thought, as was done to the previous philosophical and religious creations while working out the modern scientific cosmogonies.

19 living matter in the biosphere

The biosphere is the only area of the Earth's crust occupied by life. Only there, in the thin outer layer of our planet, is life concentrated – it contains all the organisms that are always separated from the inert matter surrounding them by an impenetrable barrier.

A living organism has never originated from inert matter. Dying, living and decaying, it gives the biosphere its atoms and incessantly takes them from it, but living matter enveloped by life always originates from living matter. Life seizes a considerable part of the atoms comprising the matter of the Earth's crust. Under its influence these atoms are in constant incessant movement. Millions of different compounds are formed by them all the time. This process has lasted without pause from the ancient Archeozoic era up till the present time, remaining unchangeable in its main features.

There is no chemical force on the Earth's surface that acts more continuously and hence is mightier in its final consequence, than living organisms taken as a whole. The more we study the chemical phenomena of the biosphere, the more we become convinced that there are no cases of them being independent from life. It has been like this throughout the whole course of geological history. Ancient archaic layers give indirect evidence of the existence of life; ancient Algonquian rocks, as well as perhaps Archeozoic ones, have retained direct imprints and obvious traces of organisms. The scientists are right in considering the Archeozoic era as rich in life as the Paleozoic, Mesozoic, and Cenozoic eras. The most ancient parts of the Earth's crust which we know and that are accessible to us, date back to that time. These layers are witnesses of the most ancient life, which has been lasting, beyond doubt, for no less than 2.109 billion years. In the course of this time the energy of the Sun could not noticeably change, and this very well corresponds to the astronomical possibilities.

20 Moreover, it becomes clear that the halt of life would inevitably cause a halt of chemical changes, if not in the Earth's crust as a whole, then at least in its surface, in the face of the Earth – the biosphere. All the minerals of the upper layers of the Earth's crust – the free aluminum and siliceous acids (clays), the carbonates (limestone and dolomite), the hydrates of ferric oxides and aluminas (brown hematites and bauxites), and many hundreds of others – are incessantly being formed in it due to life. If life stopped, their elements would soon enter new chemical groups corresponding to the new conditions, and the old bodies that are known to us would disappear. With the halt of life, the Earth's surface would be devoid of the force that can continuously give birth to new chemical compositions.

It would inevitably come to a state of chemical balance, chemical tranquillity, broken from time to time by the introduction of elements from the depths of the Earth by gaseous streams, thermal springs, or volcanic eruptions. But the new matter introduced in this way would soon take on stable forms of molecular systems inherent to the conditions of a lifeless Earth's crust; matter would undergo no more changes.

Although there are thousands of places where the matter from the deepest parts of the Earth's crust penetrates to the surface, they are scattered over the whole surface of the planet and lost in its vastness. Although being repeated sometimes, like volcanic eruptions, they are hardly noticeable in the infinity of Earth's time.

If life on Earth's surface stopped, only the slow imperceptible changes related to the Earth's tectonics would take place. They would manifest themselves not in our years and centuries, but in years and centuries of geological time. Only then they would become noticeable in the cosmic cycle, like the radioactive changes of atomic systems which manifest themselves only there.

The constantly active forces of the biosphere – the solar heat and the chemical activity of water – would hardly change the picture of the phenomenon because with the halt of life, the free oxygen would very soon disappear, and the quantity of carbon dioxide would decrease to a minimum; so the main agents in the weathering processes would disappear – those which are constantly captured by dead matter and constantly restored in the same unchanging quantity by the processes of life.

Under the thermodynamic conditions of the biosphere, water is a mighty chemical agent. This is the natural, the so called "vadose" water, rich in chemically active centers of life with organisms mainly imperceptible to the eye and changed by oxygen and carbon dioxide dissolved in it. Water devoid of life, of oxygen, of carbon dioxide, at the temperature and pressure of the Earth's surface in inert gaseous surroundings, would be an indifferent, rather inactive body.

The Earth's surface would become as unchangeable and chemically inert as the surface of the moon, the fragments of celestial bodies seized by Earth's gravity (meteorites rich in metals) or as cosmic dust permeating the celestial spaces.

21   So life is a great, constant and incessant disturber of the chemical inertness of our planet's surface. It creates not only the whole picture of our natural surroundings with its colors, forms, animal and vegetative communities and the creative activity of civilized humanity – its influence goes much deeper and penetrates into the most grand chemical processes of the Earth's crust.

There is not a single considerable balance in the Earth's crust uninfluenced by this life, which leaves ineffaceable traces on the whole chemistry of the Earth's crust. *Life therefore, is not an accidental phenomenon, exterior to the Earth's crust.* It is part of the structure and the mechanism of the terrestrial crust in which it fulfils functions of primary significance necessary for that mechanism to exist.

22   Life and living matter may be spoken of as an indivisible whole in the mechanism of the biosphere, but only a part of it immediately utilizes solar rays, namely, the green vegetation containing chlorophyll. From luminous solar rays this green vegetation produces, by means of photosynthesis, chemical compounds that are unstable in the thermodynamic field of the biosphere, which is where they will go if they leave the plant or if the plant dies.

The whole living world is directly and indissolubly connected with this green part. Animal matter and that of plants containing no chlorophyll are further developments of the same processing of chemical compounds created by green plants. Perhaps only autotrophic bacteria are not dependent on living matter, but they are at any rate genetically related to it.

So all these parts of living nature may be envisaged as the process of further transformation of the luminous solar energy into active energy of the Earth. Animals and fungi, for example, grow bodies rich in nitrogen that are even more powerful agents of chemical energy – centers of free chemical energy. When the organism dies, the compounds that are being released from it leave the thermodynamic field where they are stable and enter the biosphere – a different thermodynamic field where they decompose and release energy.

We may then regard *all living matter as a whole*; that is, the entity of all living organisms without exception as the entire and peculiar domain of transformation of the luminous solar energy into free chemical energy and accumulation of the latter in the biosphere.

23 Studying the morphology and ecology of green organisms has long shown that a green plant, both in its community and in its movement, is primarily adapted to carrying out its cosmic function; namely, capturing and transforming solar rays.

J. Wiesner, an outstanding Austrian botanist who had deeply meditated on these phenomena, stated once that light exerts much more influence on the form of plants than heat – "as if it molds them from some plastic material."

One and the same empirical generalization is presented here from two opposite points of view, and we are not able to choose between the views at present. On the one hand, the explanation is sought within an autonomous living organism that adapts itself to capturing all the luminous energy of solar rays; on the other hand, it is sought outside an organism, in solar rays that work at the green organism lighted by it as an inert mass.

It may turn out that the explanation of the phenomenon is to be sought in both, but we must leave it to the future. We must take into account the empirical observation itself, which contains, to my mind, much more than is expressed in the given explanations. Empirical observation demonstrates the existence of an indissoluble bond in the biosphere between the luminous solar radiation and the green living world of organized beings.

In the biosphere, conditions are always present to ensure the encounter of the Sun's ray with the transformer of its energy, the green plant. Such transformation of energy is sure to be normal for every solar ray and this transformation of energy may be considered to be a property of living matter, as its function in the biosphere. In case there is no transformation and a green plant cannot perform its function in the mechanism of the Earth's crust, we should try to explain the abnormality of the phenomenon.

The main conclusion of observation is that the process is absolutely automatic: when broken, it is restored without help of any other objects except solar rays and the green plant with its definite structure and definite mode of living. Only if the hindering forces are too great will the balance not be restored. The restoration of balance requires time.

24   Observation of nature provides us on all fronts with indications of the existence of this mechanism in the biosphere. Reflection easily makes us realize its grandeur and significance.

All land is covered with green vegetation. Areas free from green life are an exception and are lost in the general picture. Seen from cosmic space, the land of the Earth's surface must appear green. As continuously as the stream of solar light falls on the Earth, the green apparatus of its reception and transformation extends along the Earth's surface, including land and ocean.

Living matter in the form of a mass of organisms diffuses over the whole surface of the Earth like particles of gas; it produces a pressure on the surroundings. It avoids all the obstacles hindering its progress, or else surmounts them. In the course of time, it inevitably covers the whole planet with its layer and may be only temporarily absent when its motion is stopped or hindered by some external force. This inevitability of its ubiquity is explained by the continuous illumination of the Earth's surface by solar radiation, which creates the green living world around us.

The motion of living matter is caused by propagation of organisms; that is, by automatic increase of the quantity of their entities. It is carried out practically without interruption, and its intensity is fixed like that of solar rays falling on the Earth's surface.

In spite of the great changeability of life, it is certain that both in the sum total of organisms – living matter – and in separate organisms, reproduction and growth (i.e., the organisms' work in transforming solar energy into terrestrial chemical energy) is subject to immutable mathematical laws. Everything is calculated and proportioned to the precision, rhythm, and measure which we see in the harmonious movements of a heavenly body and which we are beginning to see in the constituents of matter and energy, such as atoms.

25   propagation of organisms and geochemical energy of living matter

The diffusion of green living matter in the biosphere through propagation is one of the most characteristic and significant manifestations of the mechanism of the Earth's crust. It is common for all kinds of living matter with or without chlorophyll; it is the most characteristic and significant indication of all life in the biosphere, its crucial difference from dead matter, a way of covering the whole biosphere with the energy of life. We see it in the surrounding nature, in the ubiquity of life, in life covering all free space of the biosphere if it meets no insuperable obstacles. The whole surface of the planet is the domain of life. If some part of it has turned out lifeless, sooner or later it will inevitably be conquered by living organisms.

In the planet's history, a measure of geological time is a short period, and we can see how during this time new organisms appear, adjusted to conditions that formerly made life impossible. The domain of life is probably extending in the course of geological time too (§119, §122); at any rate it undoubtedly covers or tries to cover all the accessible space throughout the whole course of geological history. This tendency is surely a characteristic feature of living matter itself, not a manifestation of some external force as, for example, the movement of a sand heap or a glacier under the influence of gravity.

The diffusion of life, its motion which expresses itself in the ubiquity of life, is a manifestation of its internal energy, of its chemical work. This is similar to the diffusion of a gas, which is caused not by gravity but by the motion of the particles of which the gas is comprised. Similarly, the diffusion of living matter over all the planet's surface is a manifestation of its energy, of its inevitable motion which leads to the occupation of a new place in the biosphere by the new organisms created by means of propagation. First of all, it is a manifestation of the autonomous energy of life in the biosphere. This energy is expressed in the work produced by life, in transferring chemical elements and creating new bodies from them. I shall call it *the geochemical energy of life in the biosphere.*

26 The movement of living organisms by means of propagation, which occurs with surprising and unchangeable mathematical regularity, is incessantly going on in the biosphere; through its effects it is one of the most characteristic and significant features of the biosphere's mechanism. It takes place on the Earth's surface, on land, it penetrates all the water reservoirs including the hydrosphere, it is observed throughout the troposphere, and in the form of parasites, it covers all other living beings and takes place inside living beings themselves.

Inexorably and unchangeably, it lasts without stop or delay for myriad years, and all the time it performs a great biochemical work, as it is a form of penetration of solar energy into our planet and its distribution all over the Earth's surface. We must regard it as the transfer not only of material bodies, but also of energy. In this respect, the transfer of material bodies by means of propagation is also a specific process.

It is not a simple mechanical movement of bodies along the Earth's surface independent from the environment in which they are moving. The environment the bodies are moving in creates friction by its resistance, as takes place in gravity-effected motion. The relation of this kind of motion to the environment is much deeper. This motion can take place only under the influence of gas exchange between the moving bodies and the environment in which they are moving. The more intense the gas exchange, the faster the movement; it dies away if gas exchange is not possible. Gas exchange is the *breath* of organisms; we shall see how deeply it changes propagation, how it directs it. The movement by propagation reveals its geochemical significance, its being part of the biosphere's mechanism, and the movement itself is a reflection of solar radiation. The breath itself, the gas exchange between life and its environment, is also a manifestation of the same energy of radiation.

27   Although this motion around us goes on incessantly, we do not notice it, because our eyes can cover only its final result: the beauty and the variety of forms, colors, movements and correlations that we observe in living nature. We see only forests and fields with their plant and animal life; we see rivers and seas full of life; we see the ground that is permeated by life and only seems lifeless. We see the static result, the dynamic balance of these movements; rarely can we observe them as such.

Let us consider a few examples that will reveal to us this peculiar movement creating a living nature, the movement invisible to us but crucial for living nature. Sometimes, on comparatively small areas, we observe that the life of higher plants stops. Forest fires, scorched steppes, ploughed, dug up or abandoned fields, new islands, hardened streams of lava, areas of land covered by volcanic ashes, areas freed from glaciers or water reservoirs, new soils formed on lifeless rocks by lichens and mosses – all these, as well as other forms in the endless repertoire of manifestations of life on our planet, form spots on the green cover of the planet, which are temporarily devoid of trees and grasses.

Such spots exist for a short time. Life asserts itself very soon. The green grasses and then the trees occupy the devastated or new places. This occurs partly due to penetration from outside, from the introduction of seeds by moving organisms (especially by wind) and partly from the stores of seeds that exist everywhere in the soil, which lie there in a dormant state and sometimes remain in such a state for at least centuries.

Penetration of seeds from outside is a necessary condition for population of a disturbed environment, but does not perform it. Population growth takes place due to the propagation of organisms and it depends on the geochemical energy specific for their propagation; it lasts for years, until the broken balance is restored. As we shall see, it is in complete accordance with the speed of transmission of life in the biosphere, transferring the geochemical energy of these living beings; that is, green plants.

In this case, observing attentively how waste lands are being populated, one can see the movement of the diffusion of life I am speaking about, feel its actual pressure. Considering it, one can contemplate the solar energy turned into the terrestrial geochemical energy that moves along our planet. Man feels it when he has to protect the fields or waste lands he needs from undesirable populations, spending his own energy to overcome the pressure of life. He feels it also while peering into nature, into the obscure, silent, relentless struggle for existence of the surrounding green plants. He actually sees and feels forest attacking steppe or lichen tundra choking forest while moving forward.

28   The Arthropoda, ticks, and spiders comprise the bulk of the terrestrial animal living matter. In tropical and subtropical countries the Orthoptera, ants and termites, predominate. Their propagation takes place in a peculiar way. Although they use geochemical energy (§37) of the same order as the green plants, it is not so intense.

In the termites' kingdoms there are tens or even hundreds of thousands of sexless entities, but only one organism gives progeny and actually performs the propagation; this is the queen. She lays eggs incessantly for all her life, which lasts ten or even more years. The amount of eggs, the new individuals she can produce, is counted in the billions. She produces hundreds of thousands every year. In some cases she lays 60 eggs per minute, i.e., 86,400 per day, and therefore is about as precise as a clock striking seconds.

Propagation takes place by swarming. Part of the progeny flies away with a new queen and colonizes new spaces outside the area occupied by the initial kingdom. Instinct works with mathematical precision in all cases: in preserving eggs, which are immediately carried away by working termites or in replacing the old mother with a new one in an emergency. Number also comes in everywhere with the same precision. Everything is governed by measure, by numerical regularity: the numbers of eggs, and of yearly swarms, the entities within them, the size of the kingdom's population, sizes and weights of organisms, the rate of propagation, and that of the resulting transmission of the termites' geochemical energy along the Earth's surface.

The intensity of the termites' movement along the Earth's surface resulting from their propagation can be numerically expressed: If we know the yearly number of swarms, the number of individuals within them, their size, and the number of eggs laid by the queen in a year, we can numerically express the reflection of this movement in the surroundings, or its pressure. This pressure is enormous. A man living within their habitation area knows it from the work he must do to protect the products of his existence, his nutrition, from them. Had it not been for external obstacles, mainly those created by life surrounding the termites, they could in a few years have seized and covered the whole surface of the biosphere with their kingdoms.

29 Bacteria occupy a special place among organisms. They are organized bodies of the smallest known sizes: their linear sizes are measured in 10^{-4} and even 10^{-5} cm. At the same time these organisms have the highest intensity of propagation. They multiply by division. In a day each cell redoubles several times. The bacteria with the greatest propagation rate do this work 63 to 64 times per day, on average every 22 to 23 minutes, with the same regularity as a female termite laying eggs or the planet the bacteria inhabit going around the Sun.

Bacteria live in a liquid or semi-liquid medium. The largest masses of them are observed in the hydrosphere, and large quantities are concentrated in the soil or penetrate into other organisms. Had there been no obstacles in the external surroundings, they could with extreme speed have created an incredible quantity of chemical compounds containing enormous chemical energy. The incredible speed of their propagation corresponds to this enormous chemical energy. It would take bacteria 32 hours or less to cover the Earth's surface with a thin layer, while the same surface could be conquered by propagation of plants or insects in several years, and in some individual cases in several hundred days.

The marine environment contains bacteria of almost spheroid form, the size of which, according to F. Fischer, reaches 1 cubic micron (i.e., 10^{-12} cm^3). One cubic centimeter can contain 10^{12} entities, and at a speed of propagation amounting to 63 divisions of each cell per day, a cubic centimeter can be filled by them in a few (11-13) hours if only one bacterium is initially present there. Actually bacteria do not live individually, but form colonies, and under favorable conditions they fill up one cubic centimeter ever more quickly.

The process of division always takes place at such a speed if the bacteria live under favorable conditions, especially if the temperature of the environment allows it. If the temperature falls, the speed of generative interchange decreases, and this modification can be expressed by an exact numerical formula. The bacteria are constantly breathing; that is, are dependent upon the gases dissolved in the water. It is clear that through propagation the amount of bacteria can never reach the amount of gaseous molecules in a cubic centimeter i.e., 2.706×10^{19} (Loschmidt's number). A cubic centimeter filled with water contains far fewer gaseous molecules. Here we see how propagation is limited by the phenomena of respiration, by the properties of the gaseous state of matter.

30 The example of bacteria allows us to express the motion observed in the biosphere due to propagation in a different form. Let us imagine the hypothetical period in Earth's history, when, as geologists wrongly assume, the ocean covered not two thirds of the Earth's surface, but the whole planet. E. Suess dated this "universal sea" ("panthalassa" in Greek) in the Archeozoic era. At that time it was undoubtedly populated by bacteria. Their traces have been observed in the most ancient Paleozoic layers.

The character of the minerals of the Archeozoic layers, and especially the character of their associations, proves the presence of bacteria throughout the whole Archeozoic era, in the most ancient layers of our planet accessible to geological study. If the temperature in that "universal sea" had been favorable for their life, and if there had been no obstacles to their propagation there, it would have taken the spheroid bacteria with a size of 10^{-12} cm^3 about 35.3 hours to cover that sea with a continuous film of 5.10065×10^8 km^2. The films of bacteria that appear as the result of their propagation occupy smaller but still quite considerable areas in the biosphere.

In the 1890s, Prof M. A. Yegunov had pointed out the existence of a thin film of sulfuric bacteria covering the whole area of the Black Sea.

In this case the film would be equal to the surface of the Black Sea (i.e., 411,540 km²) and would extend to the border of the oxygenated zone (200 m deep). But the studies of Prof. B. L. Isachenko, participant of the expedition of N. M. Knipovich (1926), do not confirm these data. The same phenomena, though on a smaller scale, are distinctly expressed in the dynamic balances of living organisms; for example, on the border between salt and fresh water in the Dead Lake of Kildin Island, the whole surface is always covered by a continuous film of purple bacteria (K. Deryugin, 1926).

Other microscopic, but larger organisms are constantly exhibiting examples of such phenomena. From time to time, the film formed by these ocean plankton organisms covers an area of thousands of square kilometers. Such films are formed rather quickly.

In all these cases, the geochemical energy of these processes can be expressed in the same way; that is, as the movement of this energy along the Earth's surface, with the diffusion speed directly proportional to the propagation speed of the species – in our case Fischer's bacteria. At the maximum development of the species, when the species occupies the entire Earth's surface (5.0065×10^8 km²), this energy – in a certain period of time which is different for each species – will cover also the maximum distance equal to the Earth's equator (i.e., 40,075,721 m).

While forming a film in E. Suess' "universal sea," Fischer's bacteria with a size of 10^{-12} cm³ would have developed an energy that could move around the Earth's diameter with a speed of about 33,100 cm/sec. This phenomenon can be expressed in a different form. The speed v equal to 33,100 cm/s, can be regarded as the speed of transmission of life, of geochemical energy, around the Earth;

it is equal to the average speed of bacteria rotating around it by means of propagation. In 35.3 hours bacteria could flow around the Earth by means of propagation, completing a full revolution around the planet in the "universal sea." *The speed of transmission of life* for the maximum distance accessible to it, or the quantity v, will be a constant quantity, characteristic of any homogeneous living matter. We shall use this quantity to express the geochemical energy of life.

31   This quantity, which is always different for every species or race, is on the one hand based on the propagation mechanism, and on the other hand on the limits set to the possible propagation by the volume and the properties of the planet.

The speed of transmission of life is not a simple expression of the properties of autonomous organisms or their conglomerates; it expresses their propagation with respect to the biosphere as a planetary phenomenon. Its expression necessarily includes elements of the planet such as the sizes of its surface and equator. An analogy exists with some other properties of the organism, for instance with its weight. The weight of one and the same organism on Earth and on another planet will be different, though the organism itself might remain unchanged. Similarly, its life transmission speed on Jupiter, whose surface area and equator length differ from those of the Earth, will be different, even though the organism itself did not change at all.

This specifically terrestrial character of the life transmission speed is caused by the limitations set to the propagation mechanism inherent to organisms as autonomous beings by the properties and characteristics of the Earth as a planet, of the biosphere as a planetary phenomenon.

32 The field of propagation phenomena has not attracted much attention of biologists. But several empirical generalizations have been established in this field and gone partly unnoticed by the naturalists. We are so used to these generalizations that we tend to take them for granted. Among them the following should be pointed out:

1. Propagation of all organisms is expressed as a geometric progression. This can be expressed by the formula $2^{n\Delta} = N_n$;

   n is the number of days since the beginning of the propagation;

   Δ is the progression indicator which, for unicellular organisms propagating by division, corresponds to the number of generations per day;

   Nn is the number of entities existing after n days due to propagation.

   The value Δ is characteristic of any living matter, and the formula contains no limitations, no boundaries for n, Δ or N. The process is imagined as endless, as the progression is endless.

2. This unlimited possibility of an organism's propagation is constrained in the biosphere (i.e., *the rule of the diffusion of living matter must submit to the rule of inertia.)* [The physical law of inertia says that a body will remain in motion *until acted upon by an external force.*] It is empirically established that the propagation process is retarded only by external forces: It comes

to a standstill at low temperatures, stops or weakens when there is a lack of food or respiration, and when a habitation area for newly born organisms is absent. As long ago as 1858, C. Darwin and A. Wallace put this idea in a form long known to naturalists who had thought about these phenomena, such as, Linnaeus, J. Buffon, A. Humboldt, C. Ehrenberg, and K. M. von Baer. Had it not been for external obstacles, any organism could in the course of a certain time, which differs for different organisms, cover the whole Earth by propagation and produce progeny equal to the mass of the ocean or Earth's crust.

3. The rate of propagation manifested in this way is different for each organism and is closely related to the size of the organism. *Small organisms*, which are at the same time lighter organisms, *propagate much faster than large organisms*, which are heavier.

33 These three empirical observations consider the organisms' propagation phenomena beyond time and space, or to put it more correctly, in the geometric mechanical formless homogeneous time and space. Actually, life in the form we are studying is a sheer terrestrial, planetary phenomenon inseparable from the biosphere and created and adjusted to its conditions. Life transferred to the abstract time and abstract space of mathematics is a fiction, a creation of our intellect, which does not correspond to reality. If we want to have exact scientific ideas, we must insert corrections into the abstract notions of time and space; as we see now, these corrections may dramatically change our conclusions, in which no allowance has been made for the properties of terrestrial time and space.

34   On Earth organisms live in a limited space, the size of which is the same for all of them. They live in a space of a certain structure, in a gaseous medium or in a liquid penetrated with gas. Although time seems boundless to us, the time of a certain process in a limited space, such as propagation, cannot be boundless. It has a limit, different for each organism, which depends on the organism's propagation character.

The inevitable consequence of this statement is the limitation of all quantities determining the organisms' propagation phenomena in the biosphere. There must be maximum numbers of entities that can possibly be given by different kinds of living matter. These quantities – N_{max} – must be finite and characteristic of every species and race. The life transmission speeds must be contained within exact and definite limits which can by no means be exceeded. Finally, the Δ values of the geometrical progressions of propagation also must have their limits. These limits are set up by two manifestations of the planet: 1. by its size and 2. by the physical filling of habitable space with liquids and gases, and especially by the properties of gases and by the character of the gas exchange.

35   Let us contemplate the limitation set up by the size of the planet. Everywhere we can see the influence of its size. Small water reservoirs are often covered with a continuous layer of green plants floating all over their surface. At our latitudes, these are mostly green duckweeds – different species of *Lemna*. The surface of water looks like a continuous green carpet of these plants without free spaces. The small plants are closely set together and their green plates are intertwined – the propagation process works but is hampered by an external obstacle, mostly by the absence of space. It is manifested only when free spaces of water surface appear in the green carpet as the result of various external causes, such as the death of duckweeds or their removal. These free spaces are immediately covered by propagation.

Obviously the quantity of duckweed entities that can occupy the given area is restricted and depends upon their size and the conditions for their existence. When the maximum quantity is reached, the propagation process comes to a standstill, hampered by the external insuperable obstacle. In every pond a kind of dynamic equilibrium is created, analogous to that of the evaporation of water from its surface. The resiliency of vapors and the resiliency of life are mechanically analogous.

Another well-known example is presented by green algae, a different species of *Protococcus*, the geochemical energy of which is much higher than that of duckweed. Under favorable conditions it covers tree trunks (§50) completely, without leaving free spaces. It has no further place to go; its propagation process is hindered and it starts again as soon as there is the smallest spot available for new entities of *Protococcus*. The number of entities of this algae which can cover the area of the tree is strictly determined and cannot be exceeded.

36 These ideas can be applied to living nature as a whole and to its habitation area – the surface of our planet. The highest possible manifestation of the propagation power of a living being is determined by the planet's size and is expressed by the number of entities that can occupy an area of 5.10065×10^8 km^2. This quantity is a function of the maximum possible population density of given organisms.

This density can vary greatly. For duckweed and unicellular algae of *Protococcus* it is determined only by their sizes – other organisms require much larger areas (or volumes) for their lives. The Indian elephant needs about 30 square kilometers; sheep in the Scottish highlands require about 10,000 m²; an average beehive no less than 10-15 km² (one bee needs no less than 2×10^{-4} km², or 200 m²) of average red forest of the Ukraine; 3,000–15,000 plankton entities develop best in 1,000 cm² of sea water; 25–30 cm² are enough for common cereal [individual plants]; and several (sometimes *several dozen*) square meters for the flora of our usual forest.

It is obvious that the speed of transmission of life depends upon the density of a prosperous, healthy community of individuals, on the density of living matter. I shall not stop at this significant constant value of life in the biosphere, which has not been well studied as yet. It is obvious that the highest density of a continuous covering (of the duckweed or *Protococcus* type), or the entire filling of 1 cm³ by the smallest bacteria (§29) will give us the maximum possible number of species entities in the biosphere. To get this number, it is necessary to assume the density to be equal to the square of the maximum change of the organisms, i.e., of its length and width (coefficient k_1).[7]

---

7. See my papers in Izv. Acad. Nauk, L., 1926, 272; 1927, 241; gen. sci. Paris, 1926, 661, 700

37 The limitation of propagation by the planet's size, an inevitable hindering of the process in this way, together with a deeper influence caused, as we shall see, by the green environment (§123), gives this process very peculiar and important features. *First there obviously exists a maximum distance, similar for all organisms, at which the transmission of life can expand.* It is equal to the length of the Earth's equator (i.e., 50,076,732 m).[8]

Second, for each species or race there is a maximum number of entities that cannot ever be exceeded. This maximum number is achieved when the Earth's surface is completely covered by the given species at the highest possible density of its population. From here on I shall express this number as N_{max} and call it *the stationary number of the homogeneous living matter.*[9] It is exceedingly important for estimating the geochemical effects of life. It corresponds to the maximum possible manifestation of homogeneous living matter in the biosphere, to its maximum geochemical work; the speed of reaching this number, different for every organism, is expressed by the speed v; that is, the speed of life transmission.

This speed v is connected with the stationary number by the following formula: $v = \dfrac{13{,}963.3\,\Delta}{\log N_{MAX}}$

It is obvious that at a constant life transmission speed, Δ, characterizing the intensity of propagation, must slow down on the given area or in the given volume, as the number of the created entities increases and approaches the stationary one.

---

8. $\dfrac{5.10065 \times 10^8}{365} = 13{,}963.3$, where 5.10065×10^8 is the Earth's surface. *[Note of the Editor of the 3rd edition.]*

9. That is, the matter of individuals of the same species.

38   We observe this phenomenon in our surrounding nature. Long ago
it had been noticed by naturalists, and it was emphatically stressed
about 40 years ago by the precise observer of living nature, K. Semper
(1888). Semper pointed out that in small water reservoirs at equal
conditions, the propagation of organisms decreases with the increase
of the number of entities. The stationary number is not reached or is
reached more slowly as the number of created organisms approaches
it; there is some factor, perhaps not always an external one (§43), that
regulates the process. The experiments of Pearl and his colleagues
with the Drosophila fly and with hens (1911–1912) confirm this generalization of Semper for other environments.

39   Life transmission speed can give us a clear notion of the geochemical
energy of various living organisms. Its limits vary widely and depend
on the size of the organism. In the case of bacteria – the smallest organisms – the speed approaches 33,100 cm/s; that is, to the
rate of propagation of sound waves in air. In the case of the largest
organisms – the large mammals – it equals parts of a centimeter. For
instance in the case of the Indian elephant $v = 0.09$ cm/s.

These are the extremes. Between them there is a range of life transmission speeds of other organisms. They obviously depend on the
size of the organisms, and in the simplest cases (for instance, in
that of spheroid organisms) the relation of the organism's size with
the speed v can be expressed mathematically. But there is no doubt
about the existence of a constant and universal mathematical correlation in this field, which corresponds to the old and strong empirical generalization (§32).

40   The life transmission speed gives us a clear idea of the energy of life in the biosphere, of living matter's work there, but this is not sufficient to define its energy. To do this we must take into account the mass of the organism and the energy of diffusion of its population in the biosphere as determined by the speed v. This expression:

$\frac{pv^2}{2}$, where p is the average mass of an organism for which the speed of diffusion of geochemical energy is equal to v, gives us the expression of the *kinetic geochemical energy* of living matter. In correlation to a certain area or volume of the biosphere, it can give us the expression of the chemical work that can be produced by the given species or race of organisms in the geochemical processes of this area or volume.

For a long time we have had approaches to calculating part of the geochemical energy of living matter in this way; that is, in correlation to a certain area of the biosphere, to a hectare. This is done while calculating *harvests*, the amount of useful organisms or products from a certain area. It is more fully expressed by the amount of organic matter that can be created by propagation and growth of the organisms on an area of one hectare.

Although these data are quite incomplete and explained by theory to an insufficient extent, they have already led to significant empirical generalizations. No doubt that the amount of organic matter created on a hectare is limited and most closely connected with solar radiant energy captured by a green plant. The geochemical energy gathered in such a way (i.e., by propagation of organisms) is the modified energy of the Sun.

Besides, it becomes clearer and clearer that in the case of maximum harvests, the amount of organic matter yielded by a hectare of land is of the same order as that yielded by a hectare of sea and approaches the same quantity. A hectare of land covers an insignificant layer of no more than several meters; a hectare of ocean corresponds to the layer of water where life is present, which is a number in kilometers. The similarity of the energy of life created in them obviously points at the solar irradiation from above as its source.

We shall see that this may be caused by a characteristic property of the soil, which accumulates in it a concentration of organisms possessing a tremendous geochemical energy (§155). Due to this concentration of living matter's energy, the geochemical effect of the thin layer of soil can compete with that of the vast thickness of ocean where the centers of life are scattered by the inert aquatic mass.

41 The kinetic geochemical energy of an organism, expressed by the formula: $\frac{pv^2}{2}$ taken in correlation to a hectare; to 10^8 cm², is expressed by the following formula, where: $\frac{10^8}{k}$ is the amount of organisms per hectare as they reach the stationary number (§37), and k is the coefficient of life density (§36).

$$A_1 = \frac{pv^2}{2} \cdot \frac{10^8}{k} = \frac{pv^2 N_{max}}{2 \, (5.10065 \times 10^8)}$$

It is exceedingly characteristic that the above expression is a constant quantity for the Protozoa. For all of them the expression A_1 takes the form:

$A_1 = \frac{pv^2}{2} \cdot \frac{10^8}{k} = a \, (3.51 \times 10^{12})$, (in *CGS* units) where *a* is a coefficient near to one .[10]

---

10. It corresponds to the poorly known specific weight of the Protozoa.

This formula shows that the kinetic geochemical energy of the *Protozoa* is determined by the speed v and related to the mass and the size of the organism, as well as to the rate of propagation Δ. As related to Δ, v is expressed by the following simple formula:[11]

$$v = \frac{46{,}383.93 \ \log 2\Delta}{18.70760 - \log k}$$ where the numerical coefficients, which are constant for all species of organisms, are determined by the dimensions of the planet (by its surface area and by the length of its equator, with the quantities expressed in centimeters and seconds).

The formula of speed shows that the actual limits for v and Δ cannot be explained by the dimensions of the planet alone. The maximum known value of v is equal to 33,100 cm/sec, and the maximum value of Δ is about 63–64. Can they go beyond these maxima (which, as can be derived from the given formulas, is possible even at a constant kinetic energy per hectare), or is there a condition in the biosphere that makes them stop? Such a condition exists, and it is the gas exchange of the organisms, which is inevitable and necessary for their existence, and in particular for their propagation.

42 An organism cannot exist without gas exchange, without respiration. The faster the propagation is, the more intense the respiration becomes. The intensity of life we can always judge by the degree of gas exchange. On the scale of the biosphere, we must certainly not consider the respiration of separate organisms, but the entire

---

11. $v = \dfrac{S}{t} = \dfrac{4.007572 \times 10^9}{86{,}400 \ \text{s}} = 46{,}383.93 \ cm \, / \, s$

where S is the length of the equator, t is the number of seconds in 24 hours; $18 + \ln 5 \times 10065 = 18.70762$ *(Comment of the editor of the 3rd edition)* result of the respiration; it is necessary to take into account the gas exchange, or respiration, of all organisms and to include it into the mechanism of the biosphere.

Empirical generalizations in this field have existed for a long time, but up till now they have attracted little attention and have not been taken into consideration by scientific thought. One of them points out that the gases of the biosphere and *the gases created by gas exchange of living organisms are the same.* They are the only ones existing in the biosphere: O_2, H_2O, H_2, CH_4, NH_3. This cannot be mere coincidence. Then all the free *oxygen* of the atmosphere is created on the Earth's surface *only by way of gas exchange of green organisms.* This free oxygen is the main source of the free chemical energy of the biosphere.

And in conclusion, *the quantity of the free oxygen* in the biosphere, which is equal to 1.5×10^{21} grams, is a number of the same order as the quantity of the existing and inseparably connected with it living matter, which measures 10^{20}–10^{21} grams. Both values are estimated independently from each other. The close connection of the Earth's gases with life definitely shows that the gas exchange of organisms, and first of all their respiration, must play the principal role in the gaseous regime of the biosphere; that is, it must be a *planetary phenomenon.*

43   The gas exchange or respiration determines the propagation rate; it sets up limits for the values of v and Δ. I have already mentioned (§29) that the quantity of organisms capable of existing in one cubic centimeter of medium must be smaller than the number of gas molecules in it; that is, smaller than 2.706×10^{19} (Loschmidt's number). If the value of v exceeds 33,100 cm/s, the number of entities produced by propagation of organisms smaller than bacteria (i.e., those with dimensions of orders smaller than $n\text{-}10^{-5}$ cm), will exceed 10^{19} in one cubic centimeter.

It is obvious that exchange of gas molecules between organisms and their environment being inevitable, the quantity of organisms that consume and produce gas molecules and are commensurable with the gas molecules, should grow faster the smaller their dimensions are. And finally the speed of its growth would become prodigious.

According to our present-day notions, in this way we come to physical absurdity. If the limitation of the number of entities in a cubic centimeter determines the smallest dimensions of an organism and thus sets up a maximum limit to Δ and v, then the inevitable constant correlation between the number of entities and the number of gas molecules in the given volume; that is, the *phenomena of respiration* plays a still greater part and constantly manifests itself in the processes of propagation.

Respiration must regulate this whole process on the Earth's surface, set up reciprocal correlations between numbers of organisms with different fertility, and determine, together with temperature, the value of Δ actually achievable by a certain organism; it also determines the maximum Δ responding to the dimensions of the organism, and does not allow reaching stationary numbers.

In the organic world of the biosphere, a determined fight for existence is always in progress, not only for food, but for the necessary gas as well, and the latter is more essential since it regulates propagation.

Respiration determines the maximum possible geochemical energy per hectare.

44   The effects of gas exchange and of the organisms' propagation determined by gas exchange are tremendous, even on the scale of the biosphere. Inert matter does not manifest even remote analogies to it. This is because by means of propagation, every kind of living matter can produce any amount of new living matter. The mass of the biosphere is unknown to us, but it is just a small part of the mass of Earth's crust with a thickness of about 16 kilometers and a mass of 2.0×10^{25} g. If there were no restrictions in the surroundings, masses of living matter as heavy as the entire Earth's crust could be produced by the power of propagation in the shortest, non-geological time.

The cholera vibrio and the bacterium *E. Coli* [12] can produce such a mass in 38.4 to 42 hours. The green diatomic algae *Nitrzschia putrida*, a mixotrophic organism of sea mud that feeds on decomposing organic elements and at the same time receives solar rays, can produce 2.0×10^{25} g of matter in 24.5 *days*. It is one of the most quickly propagating green organisms; maybe this is related to the fact that it receives part of its necessary organic matter in ready form. One of the most slowly propagating organisms, the elephant, can give the same mass of matter in 1300 years. But what do years and centuries mean in light of geological (i.e., planetary) time!

We must also take into account the fact that in the further course of time, new mass of the same value of 2.0×10^{14} g would appear in

---

12. Now it is *Escherichia coli. (Comment of the editor of the 3rd edition)*

much shorter periods. These figures give us an idea of the powers that manifest themselves in the processes of propagation.

45 Actually however, no organism produces such quantities. But the movements of such masses in the biosphere by force of propagation, even during one year, are not fantastic at all. These figures are not unreal. In the surrounding nature we can indeed observe manifestations of life corresponding to them. There is hardly any doubt that in a year's time life produces by way of propagation amounts of entities and their corresponding mass of matter of the order of 10^{25} g, and probably many times larger.

Thus every moment there is n x 10^{20}-n x 10^{21} g of living matter. This matter is constantly being decomposed and forms itself anew mainly by propagation, not by growth. Generations are born at intervals ranging from ten minutes to hundreds of years. Matter utilized by life is thus renewed. It exists at each moment but is only a small fraction of what is created in a year, for colossal quantities are created and decomposed even in a day.

This is a kind of dynamic equilibrium. It is supported by a mass of matter that we can hardly imagine. It is obvious that, even in a day, a tremendous mass of living matter is created and decomposed through death, birth, metabolism and growth. Who can measure the quantity of entities that are born and that perish incessantly! This task is even more difficult than counting the grains of sand on the seashore – the task of Archimedes. How can the living grains of sand be counted, grains that incessantly change their quantity in the course of time? Innumerable entities accumulate and change simultaneously in space and time. The number of them in past and present, or even in a short human period of time, undoubtedly exceeds the quantity of grains of sand on the seashore immeasurably more than by 10^{25} times!

46   green living matter

In comparison with the propagation force, with the geochemical energy of living matter, its mass present in the biosphere at every moment is rather small and constitutes about 10^{20}–10^{21} g. The existence of this mass is genetically related to the green living matter that is unique in its capability to capture the radiant energy of the Sun. Unfortunately, the present state of our knowledge does not allow us to estimate the portion of living matter constituted by the realm of green plants. It is possible to make only an approximate estimate of the quantitative aspect of the phenomenon.

One cannot assert that the mass of green living matter quantitatively dominates the whole Earth's surface, but presumably it prevails on *land*. Animal life is usually thought to quantitatively prevail in the ocean. Even if animal or heterotrophic life prevails in the total mass of living matter, its prevalence cannot be very considerable.

Can living matter be divided into two halves or almost into two halves by its mass – into the green autotrophic matter and the heterotrophic matter engendered by it? We are not able to answer this question. At any rate, there is no doubt that green living matter by itself produces mass of the same order – 10^{20}-n^{21} g that correspond to the *whole* bulk of living matter.

47   The composition of this green solar energy transformer on land is absolutely different from that in the sea. On land the herbaceous phanerogams prevail – the mass of arboreal plants constitutes a considerable part, which may be close to that of the herbaceous plants. Green algae and other cryptogams, especially the protists,[13] play a secondary part. In the ocean, unicellular microscopic green organisms prevail, and the grasses[14] like *zoster* and large algae constitute an inconsiderable portion of the mass of vegetal life. The latter are concentrated near the shores and in shallow places accessible to solar rays – their floating assemblages get lost in the boundless spaces of the ocean, like the assemblages of sargassoes in the Atlantic Ocean.

The green metaphytes prevail on land – among them the grasses propagate most quickly and possess the greatest geochemical energy. The speed of life transmission of trees is probably smaller. The green protists prevail in the ocean. The speed v for the metaphytes hardly exceeds centimeters per second – for green protists it reaches thousands of centimeters, i.e., exceeds by hundreds of times the propagation force of the metaphytes.

This phenomenon distinctly characterizes the difference between life on land and in the sea. Although green life may dominate in the sea to a lesser degree than it does on land, the total quantity of green life in the ocean exceeds its mass on land as aquatic areas prevail over land on our planet. *The green protists of the ocean are the principal transformers of the solar energy into the chemical energy of our planet.*

---

13. Protozoa

14. Macrophytes

48 The difference in energy character between the green plants of the sea and those of the land can be expressed both in exact figures and otherwise.

The formula $2^{n\Delta}=N_n$ (§32) gives us the increase of organisms per day (α). Taking *one* initial organism, we get the following (on the first day when n = 1): $2^{\Delta}-1 = \alpha$; hence, $2^{\Delta} = (\alpha+1)$ and $2^{n\Delta} = (\alpha + 1)^n$. The value α is a constant for each species – it shows the daily increase of one initial entity. The value $(\alpha + 1)^n$ obviously shows the number of entities created by propagation on the nth day; that is, $(\alpha+1)^n =N_n$.

The significance of these figures can be illustrated by the following examples. According to M. Lohmann, the average propagation of plankton (taking into account its death and being eaten) may be expressed by the value $\alpha +1$, which equals 1.2996. The same constant value for an average harvest of wheat in France equals 1.0130. These values correspond to an average ideal daily value of one organism of wheat and plankton after a day of propagation. The ratio of the quantities of the entities of plankton and wheat on the first day since the beginning of propagation is thus the following:

$$\frac{1.2996}{1.0130} = 1.2829 = \delta$$

Every day this correlation will grow in accordance with the power of δ; i.e., on the nth day of propagation it will be expressed by the value δ^n. For the twentieth day the value equals 145.9, and for the 100th day the quantity of plankton entities must exceed the quantity of wheat entities by 6.28×10^{10} times. During an annual cycle after which the development of wheat temporarily stops, this difference

(δ^{365}) reaches an astronomical figure (3.1×10^{39}). Of course, with such a difference in propagation rate, the difference in mass between an adult herbaceous plant on land weighing hundreds of grams (i.e., $n \times 10^2$ g, and a microscopic plankton organism weighing a few multimillionths of a gram (i.e., $n \times 10^{-5}$-$n \times 10^{-10}$g) disappears.

The organized green matter of the sea reaches such results due to its speed of recycling. The force given to it by solar rays could allow it to create in 50 to 70 days or even quicker a mass of matter equal to the Earth's crust in mass. The same maximum quantity of matter could be produced by terrestrial herbaceous plants in several years – for instance in five years by *Solanum nigrum*.

We must not forget that these figures cannot be quantitatively compared in order to express the roles of the green grass and the green plankton in the biosphere. In order to make such a comparison, they must be estimated for equal amounts of time from the beginning of the process; the difference increases quickly. While *Solanum nigrum* could produce 2×10^{25} grams of matter in five years, green plankton will produce amounts that are difficult to express in comprehensible figures. During the next, much shorter period needed to create the same mass of matter by herbaceous plants, green plankton would give much larger and less imaginable figures.

49 This difference between the green living matter of the land and that of the sea is not accidental. It is produced by solar rays, and depends on the difference in its effect on transparent liquid water and solid non-transparent soil. The realm of plankton, which quickly propagates; that is, which possesses an incomparably larger amount of energy in the biosphere, characterizes not only life in the ocean – it is inherent to any kind of aquatic life as compared with terrestrial life. The value δ^m can give a notion about the difference in energy between the compared kinds of living matter, but geochemically their energy manifests itself also in the mass, in the weight of the created entities. The mass of living matter that is produced is determined by the product of the quantity of its entities and their average weight p; i.e., $M = p\,(1 + \alpha)^n$.

Only in case the small organisms can really produce a greater mass of matter in the biosphere, will they, according to the general principles of energetics, gain a more advantageous position in it than larger organisms.

This is because any system achieves a stable equilibrium when its free energy equals or approaches zero; that is, when all the work possible within a system is done. We shall see that all the processes of the biosphere and those of the Earth's crust in general, as well as their general characteristics, are determined by the conditions of the equilibrium of the mechanical systems to which they can be reduced.

One such system is presented by solar rays (solar radiation) combined with the green living matter of the biosphere. This system will acquire a stable equilibrium in the biosphere as soon as solar rays do the maximum amount of work in it and create the greatest possible mass of green organisms. On land, solar rays cannot penetrate deep into its matter – all the way through they encounter bodies that are not transparent to them, and the layer of living matter produced by the rays is very thin.

Large plants such as trees and grasses have all the advantages for development in comparison with green protists. They produce a greater mass of living matter than protists, though it takes them more time to produce it. But this work of theirs is quite possible under the conditions of the land. In a short time, unicellular organisms achieve their limit of development, a stationary equilibrium (§37). In the system "solar rays-land" they present an unstable form, while the herbaceous and arboreal plants, in spite of a smaller store of energy captured by their mechanisms, can do greater work under such conditions and yield a larger quantity of living matter.

50 On many occasions we see the results of that phenomenon. In early spring when life awakens on the steppes, the steppes are very soon covered with a thin layer of unicellular algae, mainly of rapidly growing, large unicellular algae of the *Nostoc* species. This cover soon disappears in order to make way for herbaceous plants, which grow slowly and possess less geochemical energy, due to the properties of the opaque hard matter of the land. It is the grass, and not the *Nostoc* that surpasses in geochemical energy and inevitably predominates. The trees, the stones, and the soil are covered with *Protococci* which grow extremely fast. On humid days they manage to yield decigrams or grams of matter from a cell weighing millionths of a gram, and this process takes them only several hours. Their development, however, is limited even under the most advantageous conditions in countries with a humid climate. For instance in the platane copses of Holland, the trunks of the trees are constantly covered with an unbroken layer of *Protococci*, which are in equilibrium because their growth cannot proceed due to the opacity of the body bearing them. Their aquatic relatives develop in quite a different medium consisting of many meters of transparent substance.

As we shall see, the grasses and trees of the land have created their form by raising themselves up in a transparent aerial medium. Their whole image, the endless variety of their forms, shows us the same striving for maximum work, for a maximum quantity of living mass. They have found a new way to obtain this new medium of life – the troposphere.

51 The conditions in the ocean are totally different. Solar rays penetrate the aquatic masses for hundreds of meters, and unicellular algae, thanks to their geochemical energy being greater than that of green grasses and trees, have a chance of creating incomparably greater amounts of living matter than the green terrestrial plants can yield in the same period of time. The use of solar rays is exclusive here, and a stable form of life is presented by one of the smallest green organisms, not by a large plant. So, for the same reason, we observe an exceeding abundance of animal life, quickly eating up green plankton and in this way allowing it to turn the greatest part of solar radiation into living matter.

52 We see that solar rays bearing cosmic energy not only enact the mechanism of their transformation into chemical terrestrial energy, but also create the very form of transformers, which manifests themselves as living nature. Cosmic energy gives nature different shapes on land and in the sea; it also changes its structure; that is, determines the quantitative correlations existing between different autotrophic and heterotrophic organisms. These phenomena, subdued by the laws of equilibrium, are sure to be expressed in figures that we are only beginning to understand. This cosmic energy creates the pressure of life that is achieved by propagation (§27). It really demonstrates the transmission of solar energy on the Earth's surface.

We feel this pressure often in our cultural life. Changing virgin nature, freeing some areas of the land from green vegetation, Man must always oppose his efforts against the pressure of life; spend energy equal to it and endure labor. As soon as Man stops spending energy and providing the means to do this, all his constructions devoid of green vegetation will be swallowed by the mass of green organisms. Even now they capture every possible spot of the area that Man has taken away from them.

This pressure manifests itself in the ubiquity of life. We do not know regions where it does not exist at all; on the hardest rocks, on snow and ice fields, in the desert sands and on its stones, we come across manifestations of life. The immobile plants are brought there mechanically, microscopic life is constantly being conceived and halted; animals that can move by themselves come and stay to live there every now and then.

Sometimes even assemblages of life can be observed in these districts; some of their areas are rich in life, but this is not the green realm of transformers. Life in these seemingly lifeless areas is presented by birds, animals, insects, spiders, and bacteria. As compared to the green realm of plants, these areas are in fact azoic. In this sense they are related to the temporary halts of green life in the regions of our latitudes by snow cover or the winter retardation of photosynthesis. Such phenomena have existed on our planet in all geological periods.

Every place in living nature that is not occupied by life inevitably gets filled with it in the course of time, and the reasons may be quite diverse. Often a brand-new kind of flora or fauna populates the lifeless water reservoirs or newly-sprung areas of land. Under new conditions, previously unknown species and subspecies of organisms develop during geological periods. It is interesting and important that in the new structure of these organisms it is possible to recognize the structure and peculiarities of their ancestors, but in a modified form that is required by the new specific conditions of the new surroundings (L. Queneau). This morphological modification is engendered by the same geochemical energy that brings forth the pressure of life and manifests itself in the *ubiquity of life.*

At every moment of the planet's existence, the azoic regions or regions with scarce vegetation have a limited distribution, but still they exist, and on land they are more noticeable than in the hydrosphere.

It can hardly be simple chance, but we do not know why this limited manifestation of the energy of life is observed only on land; whether it is explained by the correlation of the terrestrial forces hindering life, by the force of solar rays, or by some properties of radiation unknown to us.

53   The capacity of green plants to capture cosmic energy is manifested not only by their propagation. Photosynthesis takes place mainly in minute microscopic plastids, which are smaller than the cells in which they are located. Myriads of these little green bodies are diffused throughout the plants and together give the impression of green color. While looking at any green organism, one can clearly see – both in little and in more important things – its capacity of capturing *all* accessible solar luminous radiations. The area of the green leaves of any separate vegetative organism is maximized, and they are distributed in space in such a way that not a single ray of light escapes their microscopic apparatus of energy transformation. Falling onto the Earth, the ray is always caught by an organism. This mechanism is dynamic, and its perfection surpasses mechanisms created by our will and intellect.

This determines the structure of the vegetation around us. The leaf area of forests and meadows exceeds by dozens of times the areas of planted vegetation – the meadow grasses of our latitude exceed it by 22–38 times, a field of white alfalfa by 85.5 times, a beech forest by 7.5 times, etc. These calculations do not include the outside organic world, which always fills in the spare areas existing between the large plants. In our forests, trees are replaced by green herbaceous soil vegetation, by mosses and lichens climbing the trunks and by the green algae of humid regions, which cover them and appear rapidly under any favorable conditions of warmth and humidity. In the cultivated fields covering the largest part of the land, it takes Man a great deal of labor and energy to keep the crops clean from weeds, and he very seldom manages to do this – alien green grasses spring up everywhere.

Before Man appeared in virgin nature, this structure had been especially pronounced, and up till now we can scientifically observe its remains. On the free areas of the "virgin steppe" which have remained intact in the south of Russia, a natural centuries-old equilibrium can be observed, which could be restored in one or two centuries, were it not for the effects of human intellect and will. Such a steppe of feather-grass *(Stipa capillata)* in the Cherson region was described by I. K. Pachosky: "It was like an ocean. As far as the eye could see, there was nothing but the waist-high tyrsa – a cloth of virgin vegetation often covered the surface of the ground completely, throwing shadow onto it and in this way helping retain moisture at the very surface. This allows lichens and mosses to grow between bunches of leaves or even under their cover, and to remain green in the middle of summer."

The same picture of a continuous green cover in the primeval grassy steppes of South America is described by some early observers as for instance F. Azara (1781–1801). He wrote that the vegetation was so thick there that one could see the ground only on the roads, near the brooks, and on the steep shores. These "virgin" steppes filled with green vegetation have survived only partially. They have been replaced by green fields created by Man. At our latitudes, the existence of green grasses is seasonal – their life is closely connected with the astronomical phenomenon of the Earth's revolution around the Sun.

54   In all other manifestations of vegetal life we can observe the same picture of the Earth's surface being covered with green in abundance. The thickets of tropical and subtropical forests, the taiga of moderate and northern latitudes, the savannah, the tundra – all those are untouched by Man and present different manifestations of the constant or periodically reappearing unbroken green cover of the planet. This cover is broken by Man, but it is impossible to say whether Man lessens or only redistributes the green terrestrial transformer of energy.[15]

At present, the vegetation communities and forms of separate plants are adjusted to the manifold ways of capturing solar rays, so that the latter cannot escape the microscopic chlorophyll plastids. No doubt that everywhere, excluding the constantly or temporarily azoic regions, solar rays cannot reach the terrestrial surface without going through a layer of living matter, which may exceed by a hundred times the area it would illuminate in the lifeless surrounding of dead matter.

55 Land constitutes the smaller part of the Earth's surface (29.2%). The major part of it is occupied by the sea. It is there that the main mass of living matter is concentrated. This living matter is the principal transformer of luminous radiant solar energy into a terrestrial chemical one. The green color of living matter concentrated in the sea is usually not seen – the matter is dispersed into myriad microscopic all-penetrating green unicellular algae. They float freely, sometimes concentrating and sometimes dispersing, about the whole boundless surface of the ocean, which is millions of square kilometers in size. They penetrate into every place accessible to solar rays, down to a depth of 400 meters. They are brought in by currents and go lower, but their main masses are concentrated at 20–50 m below the surface. They rise and fall in constant movement. Their propagation changes depending on the tempera-

15. Now we know it lessens it.

ture and other conditions; it increases or decreases depending on the Earth's revolution around the Sun.

One can hardly doubt that all the luminous radiation of the Sun is used here too. With the increase of depth, green algae give way to blue, brown and red algae with ultimate regularity – the red algae use the last remains of solar light which are not absorbed by the water (i.e., its blue part). As has been shown by V. Engelmann [ca. 1880], all these algae of different colors are adjusted to the maximum rate of photosynthesis under the conditions of the luminous radiation available in their surroundings.

This change of species with increase of depth is observed throughout the entire hydrosphere. Near the shores or shoals, or in such peculiar places as the Sargasso Sea of the Atlantic Ocean (connected with the geological history of the location), the unseen plankton gives way to huge floating fields or forests of algae (sometimes gigantic ones) and seaweeds, which present much more powerful laboratories of chemical energy than the largest forest massif on land. But they occupy a rather small area, which does not exceed a few percent of the total area occupied by pure plankton.

56   Eventually, the surface of our planet is from time to time covered by continuous green vegetation. The places that are always devoid of green plants, poor in life, azoic or lifeless areas, constitute hardly 5–6% of Earth's surface. Even if we take them into account, the layer of green matter covering our planet must always occupy an area that not only considerably exceeds its surface, but is also correlated with cosmic phenomena (i.e., with the Sun).

There is no doubt that even on land the average area of the green layer capturing solar rays in its maximum manifestation exceeds the surface covered with vegetation by more than a 100 times.[16] In the powerful upper layer of the world's ocean – about 400 meters deep – a green surface of the same thickness (about as thick as a plant leaf or as the layer of green protists in the soil) most probably exceeds this value by many times. Eventually, solar rays meet a continuous surface of microscopic chlorophyll transformers of its energy, which exceeds the surface of the largest planet of the Solar System (Jupiter) or almost does.

The Earth's surface is 5.1×10^8 km² whereas the surface of Jupiter is 6.3×10^{10} km². Assuming that 5% of our planet's surface is devoid of green plants and that its area catching solar rays increases by means of propagation of green vegetation by 100–500 times, the green area in its maximum manifestation will make up correspondingly 5.1×10^{10} -2.55×10^{11} km². It can hardly be doubted that these figures are not accidental and that the described mechanism is closely related to the cosmic structure of the biosphere – it must be related to the character and quantity of solar radiation. The Earth's surface constitutes a little less than 0.001 of the surface of the Sun (8.6×10^{-3}). The green area of its transforming apparatus gives figures of another order – it constitutes 0.86–4.2% of the Sun's area.

57 It is easy to see that the order of these figures corresponds to the order of the part of the solar energy that is captured by green living matter in the biosphere. In this connection it is possible to proceed from these facts while trying to explain the reason for the Earth's greenness.

---

16. Average is actually below 10 (may be 5–8). And many oceans are nutrient deserts; hence, Vernadsky's figures are greatly inflated. –*Ed.*

The solar energy caught by organisms makes up a small part of that reaching the Earth's surface, which in its turn receives only an inconsiderable fraction of the Sun's radiation. The solar energy is equal to 4×10^{30} calories per year; but the Earth, according to S. Arrhenius, receives 1.66×10^{21} calories per year. Only this cosmic energy can be taken into account considering the present degree of precision of our knowledge in this field. The radiation of all the stars that reaches Earth's surface can hardly exceed 3.1×10^{-5} % of that of the Sun, as was discovered by I. Newton. Taking into account the radiation of all the planets and the Moon, a considerable part of which is reflected, we can see that the amount of energy received by the Earth from these sources would not even reach 0.01 of the total energy received from the Sun by the Earth's surface. A large part of this energy is absorbed [or reflected] by the upper envelope of the Earth, the atmosphere, and only 40%, or 6.7×10^{20} calories per year, reach the Earth's surface and can be used by green plants.

The largest part of all this energy is used for thermal processes of the Earth's crust and is related to the thermal regime of the ocean and the atmosphere. In this regime a considerable part of it is caught by living matter as well, but we do not include it into the balance of the chemical work of life. It goes without saying that it plays an enormous role in *sustaining life* in the biosphere, but it does not manifest itself in *producing new chemical compounds*, the only measure of the chemical work of life.

For chemical work, the creation of organic compounds that are unstable in the biosphere's thermodynamic field (§89), green plants use only certain special wavelengths in the range of 670–735 nm (Dangeard and Desroche, 1910–1911). Although other rays (between

300 and 700 nm) are of some importance, their effects are comparatively small.[17]

It is due to this, and not due to imperfection of the transforming
apparatus, that a green plant uses only a small part of solar radiation reaching it. According to J. Boussingault, a cultivated green
field can catch 1% of the falling solar energy and turn it into organic
combustible matter. S. Arrhenius thinks that in an intensive culture
this figure can be increased to 2%. According to the direct observations of T. Brown and R. Escomb, it reaches 0.72% for a green leaf,
and a forest area hardly uses 0.33%.

58 These are minimum, not maximum figures. In J. Boussingault's calculations, even with S. Arrhenius' correction, only terrestrial vegetation
is taken into account. Besides, it proceeds from the assumption that
by cultivating we in fact increase the fertility of the soil and do not
just create favorable conditions for a certain crop, suppressing the
life of other plants we do not need. These calculations inevitably
fail to take into account the life of green "weeds" and microscopic
plants taking advantage of the favorable conditions of ploughed and
fertilized soil. Besides fields, there are green communities rich in life
on land, such as swamps, humid forests and humid meadows, that
exceed human crops in their amount of life (§150).

The average quantity of green vegetation per unit of sea area
(hectare) where its main mass is concentrated, probably gives
figures of the same order as its quantity per unit of land. A larger

17. We now know this is not true. Plants use *all* the visible spectrum but
especially the blue and the red wavelengths. –*Ed.*

annual quantity of living matter produced in the sea is accounted for by the higher rate of its propagation (§49). This vegetation is as quickly consumed by animals as it is produced by propagation. That is why in the areas rich in plankton and the benthos of the ocean it is possible to observe assemblages of animal life that seldom, if at all, occur on land. But no matter to what extent Arrhenius' number must be increased, even now it can be assumed that the order of the phenomenon was calculated by him correctly.

Green matter assimilates only a few percent of the solar radiant energy that reaches it – it must be more than two percent of the latter. These two or more percent are easily placed within the limits of the 0.8–4.2% of the solar surface corresponding to the green transformation area of the biosphere (§56). Some 40% of all the solar energy covering our planet reaches the plant (§57). The 2% used by plants correspond to the 0.8% of all the solar energy reaching the Earth.

59   This coincidence can be understood only by assuming that there is an apparatus in the mechanism of the biosphere that uses a certain part of solar energy *totally, completely*. The green transformation area of the Earth created by the energy of solar radiation will in this case correspond to the quantity of rays of a specific wavelength capable of doing chemical work on the Earth.

We can take the luminous surface of the quickly rotating Sun continually, illuminating the Earth for a luminous area *AB*. From this area, *from every point of it*, the luminous oscillations are incessantly being sent to the Earth's surface; only *m*% of them are rays of a wavelength that can be transformed into actual chemical energy of the biosphere with the help of green living matter. The surface of the swiftly and incessantly rotating Earth can also be replaced by an illuminated area that is equal to it in size. The diameter of the Sun being huge in comparison to the diameter of the Earth, and the Earth being situated at a great distance from the Sun, this area will be expressed in the figure by the point *T*. It will be a kind of focus of the rays coming from the luminous Sun *AB*.

The green transformation apparatus of the biosphere consists of minute particles – the chlorophyll-containing plastids. Their action is proportional to their surface, because very quickly the layer of chlorophyll matter becomes opaque for the chemical rays it transforms. Let us again replace the surface of the illuminated plastids with their area. In this case the maximum transformation of solar energy by green plants will take place if there exists a receiver of light on the Earth, the area of which is equal to *m*% of the luminous surface of the Sun or larger. In this case all the rays required by the Earth will be caught by the chlorophyll apparatus.

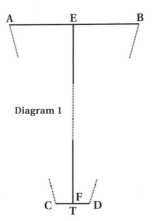

Diagram 1

In diagram 1, *CD* denotes the diameter of the circle that corresponds to 2% of the solar luminous area.[18] The whole drawing is related to the diame-

---

18. In diagram 1 the surfaces are reduced to areas – the radius of the area equal to the Sun's surface is taken as the unit. These radii are the following:

The radius of the area equal to the Sun's surface: $r = 1.3889 \times 10^6$ km^{-1}
The same for the Earth: $r_1 = 1.2741 \times 10^4$ km $= 0.00918$
The same for 2% of the Sun's surface: $r_2 = 1.9650 \times 10^5$ km $= 0.14148$
The same for 0.8% of the Sun's surface: $r_3 = 1.2425 \times 10^5$ km $= 0.08947$
The average distance of the Earth from the Sun expressed in the same scale will make up 1.495×10^8

ters of the circles, the areas of which correspond to the luminous surface of the Sun and to the illuminated surface of the Earth (T and CD). There are probably numeric correlations between solar radiation, its character (the percentage of the rays m), and the area of green vegetation (and azoic gaps?), but they are unknown to us at present.

The structure of the biosphere cannot but deeply depend on its cosmic character.

60 Living matter always retains a part of the received radiant energy within itself, within the living organisms. This value corresponds to the quantity of organisms. As we shall see, everything indicates that the quantity of life on the Earth's surface not only suffers little change over short periods of time, but also remains unchangeable or almost unchangeable[19] throughout geologic periods (from the Archeozoic era up until now).

The immediate dependence of the quantity of life on the radiant energy of the Sun makes this empirical generalization especially important because it connects that quantity with such a value as solar radiation, which is seen as constant throughout geologic time; that is, the time of the existence of the Solar System in its present form. The dependence of the main part of life, the green living matter, on solar radiation of a certain wavelength, and the mechanism of the biosphere that is being unveiled to us and is connected with *the complete use of these rays* by green vegetation, confirm once more the constancy of the quantity of living matter in the biosphere.

---

19. As with in any equilibrium, it fluctuates around the static state.

61   The quantity of energy present in the form of living matter each second can be calculated. According to the calculations of S. Arrhenius, the green vegetation in the form of its combustible compounds (i.e., organic compounds) contains 0.024% of all the solar energy reaching the biosphere – 1.6×10^{17} calories.

This is a huge number on the planetary scale, but it seems quite underestimated to me. In another paper,[20] I tried to make it clear that the number of Arrhenius was to be increased by at least 10 times or even more. Probably more than 0.25$ of the solar energy received by the Earth is stored in living matter in the form of compounds existing in a special thermodynamic field that is different from the thermodynamic field of the biosphere's non-living matter.

In spite of the enormous quantities of matter passing through organisms during their life, and the even greater quantities of, for instance, free oxygen that they produce (about 1.5×10^{21} g), the annual energy effect of life is expressed in smaller figures than the beings it produces because these are constantly restored by propagation or dying. As was already mentioned (§45), the mass of elements moving annually exceed the weight of the Earth's crust of about 16 km in thickness by many times and is expressed in numbers of the order of 10^{25} g.

As far as we can calculate at present, the annual energy contribution of living matter to the biosphere is not much larger than the energy retained by living matter in its thermodynamic field for hundreds of millions of years. It contains no less than 1×10^{18} calories in the form of combustible compounds, and it annually uses no less than

[20.] See *The Essays on Geochemistry.*

2% of the energy received by the surface for creating the new and restoring the spent compounds. This makes up no less than 1.5×10^{19} calories. If this number proves to be larger in the course of further investigation, the order of 10^{19} will hardly change. As the quantity of living matter remains unchangeable throughout the whole course of geological time, the energy related to its combustible part may be considered constantly inherent to life. In this case $n \times 10^{19}$ calories will be the number expressing the energy that is annually transferred into the biosphere by living matter.

62   some remarks on living matter in the mechanism of the biosphere

Green living matter, in spite of its importance, does not cover all the principal manifestations of life in the biosphere. The chemistry of the biosphere is filled with the phenomena of life, with the cosmic energy caught by the latter, and cannot be understood even in its general features without finding out about the place of living matter in the mechanism of the biosphere. Besides, it is only partly related to the realm of the green plant.

This mechanism is not sufficiently known to us, but even now we can indicate some of its regularities that we must consider to be empirical generalizations. In the future, the picture of the phenomenon will undoubtedly look different to us, but imperfect as it is, we must take it into account on many occasions even now. I shall briefly dwell upon some of its details, which seem essential to me.

In the history of the chemical composition of living matter, a peculiarity has long been noticed, which regulates all its geochemical history in the biosphere. It was first mentioned by the profound Russian natural scientist, K. M. von Baer; he expressed it for carbon. Later the same was noticed for nitrogen, and it can be fully applied to the entire geochemical history. It is the law of *economy* that is involved in the use of simple chemical substances by living matter once they are included into its composition.

The economy involved in the use of chemical elements necessary for life by living matter manifests itself in different ways. On the one hand, it is observed within the limits of the organism itself. An element once included into it passes through a long series of states within it, enters a series of compounds before it leaves the organism and is lost from it. At the same time, the organism introduces into its system only the vitally important quantities of elements and avoids any surplus. But this is only one aspect of the phenomenon noticed by K. M. von Baer. It is probably related to the autonomy of the organism and to its systems of equilibrium, which reach the state of stability possessing the smallest free energy.

This peculiarity of the geological history of organisms is even more pronounced in the mass of living matter. The law of economy manifests itself in countless biological phenomena. Atoms absorbed by some form of life, caught by a singular torrent of life, hardly, if at all, return to the inert matter of the biosphere. Organisms devouring each other, parasites, symbiotic organisms and saprophytes, immediately turn the excreted remains of life back into the living form of matter. A large part of these remains is actually alive – full of microscopic life forms. The new generations produced by propagation, all these various, countless organisms, absorb atoms from the changing surroundings and retain them in the whirlpools of life by carrying them from one to another.

All this has been taking place throughout the whole cycle of life for hundreds of millions of years. Some of the atoms of the unchanging layer of life, the energy of which remains at a constant level of the order of 10^{18} calories, never leave the life cycle. Using the metaphor of K. M. von Baer: life is economical in spending its captured matter and gives it back reluctantly and with difficulty. Normally it does not allow it back for long, if at all.

63 Taking into account the "law of economy," it is possible to speak about the atoms remaining within the limits of living matter during geological periods; these atoms are constantly moving and migrating but do not go back to inert matter. This empirical generalization, showing us an absolutely unexpected and unique picture, makes it necessary to try to find out both its consequences and causes. Now this can be done only hypothetically. And, first of all, this generalization makes us face a question that is new to science, but which has existed in other fields (in philosophical and theological speculations). Are the atoms thus held by living matter the same as we observe in inert matter? Or are there different isotopes, special mixtures among them? This question, which has become urgent, can find its answer only in further experience.

64 One of the most important manifestations of life in the biosphere (§42) is the gas exchange of organisms with their gaseous surroundings. Part of this gas exchange was correctly understood by L. Lavoisier as *burning*. In this way the atoms of carbon, hydrogen and oxygen are constantly leaving and entering the whirlwinds of life in great masses.

It is quite possible that burning within an organism does not touch the main substratum of life – the protoplasm. It is possible that in a living organism the atoms of carbon going to the atmosphere or water in the form of carbon dioxide originate from the alien matter entering the organism; that is, from food, and not from the matter constituting the carbon framework of an organism. In this case, only the protoplasmic foundation of life and its derivatives gather the atoms which are held by living matter without leaving it.

Now it is necessary to revise the notions about the character of the exchange or movement of atoms within an organism, about the stability of protoplasm – notions that were put out by C. Bernard and since then have appeared in science more than once. Maybe there is some connection between the ideas of C. Bernard, the generalizations of K. M. von Baer concerning the economy of life, and the fact stipulated by geochemistry that the quantity of living matter in the biosphere is constant. All these ideas may concern one and the same phenomenon; that is, the constancy of the mass of protoplasm derivatives in the biosphere throughout geologic eras.[21]

65 Studying the phenomena of life on the scale of the biosphere also gives us more definite indications of the close connection between

---

21. According to present notions, an organism receives energy by way of oxidation of its own (cellular) elements, and the elements of food replenish their decrease. Consequently, the tissue composition of long-lived organisms renews itself many times during their lives. For instance, the cycle of erythrocytes renewal of man lasts about 3 months; for the cells of the skin and mucous membrane this period is even shorter. The same happens to the boundary of cells constituting the cortex of perennial plants. *[Comment of the editor of the 3rd edition.]*

life and the biosphere and shows us that the phenomena of life must be treated as part of the biosphere's mechanism. Furthermore, functions performed by living matter in this complicated but quite orderly mechanism of the biosphere have an essential influence on the character and structure of living beings.

One of the most important phenomena of this kind is *the gas exchange of organisms, or their respiration.* Organisms are responsible for a major part of the gas exchange of the planet. In 1844, J. Dumas and J. Boussingault pointed out, in their outstanding lecture in Paris, that living matter can be considered as an appendage of the atmosphere. During its lifetime, it creates bodies from the gases of the atmosphere – oxygen, carbon dioxide, water, compounds of nitrogen and sulfur – and then it converts them into combustible solids and liquids by involving the cosmic energy of the Sun. After the body's death and during its lifetime, its gaseous elements are restored to the atmosphere.

This notion is undoubtedly true to reality. The genetic connection of life with the gases of the biosphere is immense. It is even deeper than it seems to be at first sight. The gases of the biosphere are always genetically connected with living matter and determine the basic chemical composition of the atmosphere. I have already pointed out this phenomenon while speaking about the significance of this exchange for creating and determining the propagation of organisms; that is, the manifestation of their geochemical energy (§42).

All the gases of the atmosphere, such as free oxygen and carbon dioxide, maintain a dynamic equilibrium, a constant exchange with living matter. The gases lost by living matter return there immediately;

they often go out and back into the organism almost at one and the same moment. The gaseous current of the biosphere is thus closely connected with photosynthesis, the cosmic source of energy.

66   After the destruction of an organism, the greater part of the atoms immediately enters living matter again. Only a small part, an insignificant percentage of its mass, forsakes the processes of life for a long period. This small proportion is not accidental. In all probability it is constant and unchangeable for each element. It returns into living matter in a different way after thousands and millions of years. During this intermediate time, the substance that has left the living matter plays a great role in the history of the biosphere and even in that of the Earth's crust, because a large portion of these atoms leaves *the limits of the biosphere for a long time.*

Here we are observing a new process, namely, *the slow penetration of the Sun's radiant energy, which has reached the Earth, into the planet.* Through this process living matter transforms the biosphere and also the Earth's crust. Unceasingly it relinquishes certain elements, creating enormous masses of vadose[22] minerals that do not exist separate from it. Or it disperses the fine powder of its debris throughout the dead matter of the biosphere. On the other hand, living matter fired with cosmic energy breaks the structure of the compounds formed independently of its immediate influence (§140 and further).

The Earth's crust accessible to our observation is changed by these means. Throughout geological time, modified radiant cosmic

---

22. Vadose, in geology, refers to anything located above the water table. –*Ed.*

energy penetrates more and more into the interior of the planet
through this action of living matter. The vadose minerals, turning
into phreatic[23] forms of molecular systems, are the means of this
transportation. The non-living matter of the biosphere is created
largely through the agency of life, and thus we see that the ideas
about its geological significance as expressed by the philosophers
and natural scientists Ocken, Steffens, and Lamarck of the early
nineteenth century, were closer to the truth than those of the geologists of subsequent generations.

It is characteristic that such an influence on the matter of the
biosphere, especially on the creation of vadose mineral deposits,
is mediated by the action of aqueous organisms. The constant
movement of water reservoirs throughout geological time distributes the aggregations of free chemical energy of cosmic origins over
the whole planet. All these phenomena probably have a character
of stable equilibrium, and the masses of matter taking part in it
undergo few changes after the energy of the Sun reaches the Earth
and creates them.

67   Eventually, a great amount of the matter of the outer envelope of
the Earth – the biosphere – is caught and gathered by the living
organisms that are modified by the cosmic energy of the Sun. The
weight of the biosphere must be of the order of 10^{24} g. But the active
living matter itself, which transfers the solar energy, constitutes in
general not more than 1% of the mass of this outer envelope of the

---

23. *Phreatic*, in Geology, refers to ground water or explosive volcanic activity
involving steam derived from ground water. – *Ed.*

Earth – perhaps it is even a fraction of a percent. Nevertheless, in some places it dominates the non-living matter, and in a thin layer, for instance in the soil, it can constitute much more than 25% of the mass.

So the appearance and formation of living matter on our planet is evidently a phenomenon of cosmic character that is clearly characterized by the absence of abiogenesis; that is, by the fact that throughout the entire geologic history, a living organism always appears from a living organism, all organisms are genetically inter-connected, and we never observe a solar ray being caught and its solar energy being converted into chemical energy except by an already existing organism.

How was this special mechanism of the terrestrial crust formed, this matter of the biosphere penetrated with life, this mechanism that functions ceaselessly throughout myriad years? We do not know. It is a mystery just as life itself is a mystery in the general scheme of our present knowledge.

the domain of life

68　the biosphere: the earth's envelope

The importance of life in the structure of the Earth's crust has penetrated slowly into the scientific mind and has not yet been fully appreciated. It was as late as 1875 that one of the most outstanding geologists of the last century – E. Suess of the University of Vienna – introduced the idea of the biosphere as a special layer of the Earth's crust; a layer permeated by life. Thus he completed the idea of the ubiquity of life and the continuity of its manifestations on the Earth's surface.

Having introduced the notion of a special layer of the Earth's crust conditioned by life, E. Suess actually expressed a novel and very broad empirical generalization, without seeing all its possible consequences. Only at present is this generalization getting its explanation thanks to new scientific achievements, which did not exist in his time.

69　The biosphere forms the upper envelope, or geosphere, of one of the vast concentric parts of our planet's structure – the Earth's crust.

The physical and chemical properties of our planet change continuously according to their distance from the center. Investigation can confirm that they are identical for concentric segments.

It is possible to differentiate two forms within this structure: on the one hand the large concentric parts of the planet, or concenters, and on the other hand, the smaller subdivisions called the Earth's *envelopes* or *geospheres*.[24] Matter in each of these regions of the Earth's crust is apparently separated from matter in any other region, and if this matter passes from one part to the next, this process takes place exceedinly slowly or periodically, and is not a fact of its contemporary geologic history. Each region apparently presents a closed mechanical system independent of the other.

For hundreds of millions, if not billions of years, the Earth has generally retained the same temperature conditions. The thought is inevitable that during this time the conditions of stable equilibrium of matter and energy have been set up in it, as there was no external (for the mechanical systems constituting it) flow of active energy. Most probably, the lesser the flow of external energy, the more perfect the equilibrium of the closed parts of the Earth. Such parts are at least three: 1. the core of the planet, 2. the intermediary layer, sometimes called "sima" (according to Suess), and 3. the Earth's crust.

70   *The core of the Earth* has a chemical composition that is absolutely different from the Earth's crust on which we are situated. The matter of the core is possibly in a special gaseous state (of a post-critical gas), but our notions about the physical state of the matter of the deepest parts of the planet that is subject to the pressure of many dozens if not hundreds of atmospheres, are nothing more than guesses. It is possible to assume the existence of heavy elements or

---

24. The word "geosphere" is used by many geologists and geographers in the mentioned meaning. D. Murray (1910) seems to have introduced this expression. He based himself on the ideas of E. Suess.

their simple compounds in the core of the Earth in solid, viscous or gaseous states. The temperature there may be thousands of degrees almost as low as the temperature of cosmic space. Usually the possibility of the last assumption being true is ignored, and this fact distorts the estimation of the limits of our ignorance. The different and unusual composition of the core in relation to the Earth's crust is a consequence of the large specific gravity of the planet (5.7), as compared to the specific mass of the upper envelopes of the Earth's crust (2.7). The specific gravity of the core is at least 8, or even 10 and more.[25] Some assume (and this assumption is not absurd) that it consists of metallic iron and its metallic compounds.

There is no doubt that at a depth close to 2900 km below sea level, the mechanical properties of the planet's matter change greatly. This fact has been strongly confirmed by observation of earthquakes. Such a change in the properties of matter is often explained by the hypothesis that, at that depth, seismic waves penetrate into another shell. In this case that depth would correspond to the surface of the metallic core. But it is also possible to assume that the depth of these boundaries is not so great (1200 or 1600 km) and corresponds to other discontinuities observed in the course of seismic waves.

71 New data in this field will be received much sooner than has recently been considered. If we compare the results of petrogenic investigations with the results of seismic observations, we shall notice that the volume of the planet's structure with rocks

---

25. According to A. Sieberg (1923) who based his ideas on those of E. Wiechert, the specific mass of the core is 9.1.

containing silicates and aluminosilicates is much larger than it was earlier thought to be. This becomes clear mainly through the remarkable observations of the Croatian scientists A. and S. Mohorovicic, father and son. They have lately drawn attention to this fact, and their works present an undoubted achievement in comparison with their predecessors' investigations.

72   Now let us try to state some essential peculiarities of the second concentric shell of the Earth, which was called "sima" by E. Suess.[26] He supposed that it was characterized by a preponderance of the atoms of silicon, magnesium, and oxygen. This region is characterized first of all by its thickness; it occupies many hundreds of kilometers – maybe more than a thousand kilometers. Second, it is characterized by the fact that it contains five chemical elements (silicon, magnesium, oxygen, iron, and aluminum), which play a very important part in it. The increase of iron with depth is apparently explained by its heavier mass.

It is possible that rocks similar to the main rocks of the Earth's crust – the third region – are also of great importance in the structure of the sima region. In the opinion of many geologists and geophysicists, the mechanical properties of these rocks resemble eclogites.[27]

---

26. Now it is called the mantle. *[Comment of the editor of the 3rd edition]*

27. Eclogite is a rock consisting of a granular aggregate of green pyroxene and red garnet, often containing other minerals as well. – *Ed.*

73 The upper border of the sima region in the Earth's crust has an average thickness of a little less than 60 km. This has been stated rather exactly by various investigations independent of one another; on the one hand it was done by investigating earthquakes and on the other hand by measuring gravity. The sima region is separated from the Earth's crust by an isostatic surface. This surface shows a peculiarity of the sima region that makes it crucially different from the Earth's crust; its substance is homogeneous in all its concentric layers. The physical and chemical properties of the sima change concentrically according to the distance of the observed points from the center of the planet. The substance of the Earth's crust on the contrary diverge within one and the same concentric layer at the same distance from the center of the planet. Under such conditions, any considerable exchange between the substances of the sima and the Earth's crust cannot take place.

74 These data make us first of all put aside any ideas about the sima as a region rich in free energy. Its energy in relation to the phenomena we are investigating can only be potential, and its manifestations have never reached the surface of the Earth. It has not reached it throughout geologic time; that is, for hundreds of millions of years. This statement can be accepted as an empirical generalization confirmed by all the logical force of geologic observations.

In other words, there are no data to deny that the sima is in a state of chemical indifference, of stable equilibrium, complete and unchangeable throughout the whole length of geologic history.

This possibility is indicated by the fact that we do not know a single case of matter flowing from the deeper parts of the planet situated outside the limits of the Earth's crust. It is also indicated by the absence of phenomena in which the assumed free energy of the sima, for instance its high temperature, manifests itself. The free energy which penetrates to the Earth's surface from the depths – the heat – originates not from the sima, but from the atomic energy of radioactive chemical elements.[28] These elements are probably concentrated in the Earth's crust under conditions that allow the manifestation of their energy in a form capable of performing work.

75 Apart from earthquakes, it is the distribution of gravity that gives us the chance to penetrate deeply into the planet. Its essential feature is the fact that it is connected with the very peculiar and definite structure of the upper part of our planet. The distribution of gravity shows that the large parts of the crust possessing different specific gravities (from 1 for water to 3.3 for basic rocks) are all concentrated in the upper layer of the planet. They are situated in such a way that in the vertical dimension the light areas are compensated by heavier ones, and that at a certain depth – *on the isostatic surface* – an immutable equilibrium is established. Beneath that surface each layer of the planet has one and the same specific mass throughout a whole layer.

It can be concluded, therefore, that beneath the isostatic surface there is no chance for mechanical infringements and chemical differences in layers of the same depth – complete equilibrium of

---

28. Here, Vernadsky uses "atomic energy" in its modern sense. –*Ed.*

matter and energy must exist there. In view of this, it is convenient to take the isostatic surface as the lower border of the Earth's crust and the upper border of the sima. It establishes a very important peculiarity of the planet; it separates the *field of changes* from the field of immutable stable equilibria.

In the first essay we have seen that the face of the planet, the biosphere, the upper envelope of these changes, receives the energy that causes changes within it from its cosmic surroundings, from the Sun. We know, and we shall further see, that it possesses mechanisms that transfer active solar energy into the depth of the biosphere. But the Earth's crust possesses another source of free energy: radioactive matter, which causes even more powerful infringements on its stable equilibria

We do not know whether radioactive atoms reach the sima, but it seems obvious that the quantity of radioactive elements in the sima cannot be of the same order as in the Earth's crust, as otherwise the thermodynamic properties of the planet would have been quite different; apparently, the radioactive elements – the sources of the free energy of our planet – do not enter the sima, or soon disappear within it.

76 Our notions about the physical state of the sima are very incomplete. The temperature of this shell is apparently not very high, and the extraordinary state of its matter is caused mainly by the effects of an enormous pressure. The mechanical peculiarities of this substance, which goes at least 2000 km deep, are distinctly different from all the states we are used to, but they have much in common with those of the solid state (S. Mohorovicic, 1921).

The pressure at these depths is so great that it exceeds our imagination and upsets our experimentally grounded notions about the three states of matter: the solid, the liquid, and the gaseous states. At the upper border of the sima where the pressure reaches 20 thousand atmospheres, any difference between the solid, the liquid, and the gaseous states with their usual characteristic parameters ceases to exist, as follows from the experiments of P. W. Bridgman (1925).

Obviously, such a substance cannot have a crystalline structure. It may have a glassy structure or a metallic structure under great pressure; these are the most satisfactory ideas we can have about it. The layers of this region are quite homogeneous, and the greater the pressure, the more they change with increasing depth.

77 The exact depth of the isostatic surface is not known. At first it was ascribed to a depth of 110–120 km. The latest estimate gives smaller figures of about 60 and 90 km. Its level probably has different values in different places, and its form inevitably undergoes slow changes under the influence of the sources of free energy that are situated in the Earth's crust – this is the process we call geologic changes.

Above the isostatic surface there is the region of the planet called the *Earth's crust*, a name that originates from the old geologic hypotheses that assumed that, in the course of the geologic studies of the Earth's crust, we came across the traces and remains of the frozen crust of a planet that had once been in an incandescent state. It was connected with scientific cosmogonic hypotheses about Earth's past, the deepest expression of which was the hypothesis of P. Laplace. The idea had acquired a wide popularity among scientists who tended to overestimate its scientific value, but gradually it became clear that none of the accessible layers show us traces of such a frozen crust, and that there are no indications of the hypothetical incandescent past of our planet. But the term "the Earth's crust" has been retained in science, having acquired a different meaning.

78 In this *Crust* we differentiate a series of *envelopes,* which are distributed in it as concentric circles, although on the whole the bordering surfaces are not spherical. Each concentric envelope is characterized by its own, largely independent and closed systems of dynamic equilibrium, both chemical and physical. The delimitation of separate points is sometimes difficult, apparently due to large gaps in our knowledge.

The delimitation can be done more exactly for the upper parts of the solid phase of the planet and for the lower parts of the gaseous one. Chemical compounds reach or reached 16-20 km below, and 10–20 km above the Earth's surface. Investigation of the geologic composition of the Earth shows that the deepest massive rocks known to us had formed no lower than at the mentioned depths. The thickness of 16 km corresponds to the thickness of sedimentary and metamorphic rocks. It is possible to think that the chemical composition of the upper 16-20 km is caused by the same geologic processes we are studying now. This composition is on the whole well-known to us.

Beyond these limits our knowledge becomes much less exact. This is not only because we cannot find out exactly about the substance reaching us from there, but also because the *states of substance* within these limits of high and low pressures are not quite clear to us in spite of the considerable progress of experimental sciences. But we are on solid ground – the development of our knowledge proceeds slowly but persistently. And it is obvious that our old notions of the Earth's crust are beginning to be crucially revised.

79   From this standpoint it is necessary to outline a few phenomena that are essential for understanding the composition of the Earth's crust.

First, in the high layers of the gaseous envelope of the planet, the state of the substance is distinctly different from what we are accustomed to seeing around us. We are possibly dealing here (above 80-100 km) with a region of the planet that is different from the Earth's crust. In this rarefied medium, great stores of free energy are concentrated in the form of electrons and ions; the role of this energy in the history of the planet is not yet clear to us.[29]

It seems almost beyond doubt, at present, that the incandescent state of the inner layers of the planet, which was considered to manifest itself in volcanic rocks pouring to the surface, does not exist. It would be necessary to assume the existence of large or small areas of magma – i.e., the viscous, liquid, hot (600–1000 °C) silicate melt filled with gases – among the prevailing solid or semi-solid viscous hot envelope. Nothing suggests that volumes of magma permeated

---

29. Only later will radiation zones be discovered in the magnetosphere of Earth, with the help of satellites. *[Comment of the editor of the 3rd edition.]*

the entire Earth's crust and that the temperature of the crust was as high as the temperature of those hot melts full of gas.[30]

80 Although the structure of the deepest part of the Earth's crust still has many mysteries, the progress of science in this field during the last years has led to outstanding achievements. In all probability, the Earth's crust consists of acidic and basic rocks that can be observed on the surface as well. The acidic rocks such as granites and granodiorites, are situated under continents, and their thickness reaches 15 km or a little less. Basic rocks predominate at greater depths. Under the hydrosphere they approach the Earth's surface; these rocks are poorer in free energy, in radioactive chemical elements.

The existence of at least three envelopes beneath the Earth's surface must be assumed. One of them, the upper envelope, corresponds to the acidic rocks (the granite envelope). It stops at a depth of 9-15 km below the surface and is relatively rich in radioactive elements. About 34 km below the surface the properties of the substance undergo another considerable change (H. Jeffreys, S. Mohorovicic), which probably shows the lower border of a crystalline substance. At the same time it is the upper border of the glassy envelope of R. Daly (1923). Still deeper there are basic rocks and partially acidic rocks in a state close to glass, a state that is unfamiliar to us.

The second sharp change occurs at a depth of about 60 km from the Earth's surface; apparently, it is due to the appearance of heavy rocks, the influence of which expresses itself in seismic phenom-

30. The dispersed character of the incandescent magma and the absence of a continuous incandescent envelope are completely confirmed by geophysical studies.

ena. These may be eclogites[31] with a density of no less than 3.3-3.4. Here we enter the region of the sima; the specific weight of the rocks continues to increase and reaches 4.3–4.4 at its border (L. Adams and E. Williamson, 1925). These short notes give only a very general impression of the complexity of the phenomenon.

81   The existence of the terrestrial envelopes was discovered empirically and took a long time. Some of them – for instance the atmosphere – were discovered centuries ago, and the fact of their existence became customary in everyday life. But only as late as the end of the nineteenth and the beginning of the twentieth centuries were their characteristic features comprehended, and the understanding of their role in the structure of the Earth's crust has not yet entered the general scientific mentality.

The differentiation of the envelopes is closely connected with the chemistry of the Earth's crust, and their existence is the result of the fact that all the chemical processes of Earth's crust are subject to the same mechanical laws of equilibrium. That is why in the extremely complex chemical structure of the Earth's crust, it is easy to notice and identify common features that allow us to empirically differentiate their basic states in the complex of natural phenomena, and to classify the complex systems of dynamic equilibria to which Earth's envelopes correspond in such a simplified form.

---

31. Judging by the stucture of these formations, which is obviously not crystalline, these cannot be eclogites; they only correspond to eclogites in their specific gravity. The eclogites of the upper layers of the Earth's crust correspond to the deepest parts of the Earth's crust that can be studied visually.

The laws of equilibrium in general mathematical form were discovered by J. W. Gibbs in 1884–1887, who reduced them to the correlations existing between the independent variable characteristics of chemical and physical processes, such as temperature, pressure, physical state, and chemical composition of the bodies taking part in the process. All the Earth's envelopes (geospheres), distinguished purely by empirical methods, may be characterized by a few variables included into the equilibria investigated by Gibbs.

Thus it is possible to distinguish thermodynamic envelopes determined by the values of temperature and pressure, phase envelopes characterized by the physical state (solid, liquid etc.) of the bodies they contain, and finally, chemical envelopes, which are characterized by their chemical composition.

Only the envelope identified by E. Suess – the biosphere – remains aside. All its reactions are subject to the laws of equilibria, but they include a new feature, a new independent variable ignored by Gibbs.

82   Usually the independent variables taken into account while studying the non-homogeneous equilibria in chemical laboratories, such as temperature, pressure, state, and composition of matter, do not embrace all their forms. The equilibria mathematically studied by Gibbs were electrodynamic. In the natural terrestrial equilibria, different surface forces are of great significance. In chemistry, considerable attention was drawn to the phenomena of photosynthesis, where the independent variable is solar radiant luminous energy. In the phenomena of crystallization, we take into account the vectorial crystalline energies: the inner energy – for instance in the formation of crystal twins – and the surface energy in all crystal formations.

Living organisms introduce luminous solar energy into the physical and chemical processes of the Earth's crust. Nevertheless, they differ essentially and distinctly from the other independent variables of the biosphere. They change the course of its equilibrium like other variables, but unlike them they are special autonomous formations, a sort of special secondary system of dynamic equilibria in the primary thermodynamic field of the biosphere.

The autonomy of living organisms expresses the fact that their customary thermodynamic field possesses quite different parameters from those observed in the biosphere. In this relation, some organisms retain their temperature in a medium of different (often distinctly different) temperature and have their own inner pressure. They are detached within the biosphere, and its thermodynamic field is important for them only because it conditions where the autonomous systems can exist, but not their internal field.

From the chemical point of view, their autonomy is clearly expressed by the fact that the chemical compounds developing in them usually cannot appear outside them in the common conditions of the non-living surroundings of the biosphere. As they get into such surroundings, they inevitably prove unstable and decompose, entering other bodies. Being sources of free energy, they break the equilibrium of the surroundings in this way.

In living matter they often appear under conditions distinctly different from those we observe in the biosphere. In the latter for instance, decomposition of carbon dioxide and water molecules (i.e., one of the main biochemical processes) can never be observed.

In the conditions of our planet, it can only take place in the deep parts of the magmosphere outside the biosphere. In our laboratories we can produce it only at high temperatures that do not exist in the biosphere.

It is clear that the thermodynamic field of living matter is absolutely different from the thermodynamic field of the biosphere, no matter how that difference is explained. Empirically, living organisms can be described as specific thermodynamic fields: foreign to the biosphere, delimited within it, and having inconsiderable dimensions in comparison. They transfer energy from solar rays and are created by the latter in the biosphere. Their volumes range between the limits of $n \times 10^{-15}$ and $n \times 10^{-12}$ cubic centimeters.

No matter how we explain their existence and their formation in the biosphere, the fact is that chemical equilibria change in their presence, while the general laws of equilibrium are not broken. Living beings taken as a whole; that is, living matter corresponding to them, can be considered as a special form of the independent variables of the planet's energy field.

83 This influence of living beings is most closely related to their nutrition, respiration, and their destruction and death; that is, to the life processes during which chemical elements enter and leave them. There is no doubt that on entering a living organism chemical elements find themselves in surroundings that have no analogs anywhere on our planet. We express this phenomenon by saying that, on entering organisms, chemical elements exhibit a new form of *existence*. Their entire history in this form of existence is different from that in other parts of the planet. This difference is obviously caused by a deep change of the atomic systems within living matter. There are grounds to consider that chemical elements within it do not form the usual isotope mixtures. This must be proved by experiment.

Earlier scientists – and some of the modern ones too – connected the peculiarity of a chemical element's history within living matter to the enormous prevalence in it of the dispersed state of the elements' *compounds*: their *colloidal systems*. But the same kind of colloidal systems can be observed in the biosphere in other cases and obviously have nothing to do with living organisms. According to our present-day notions, the dispersed systems (colloids) are always connected with *molecules* and never with atoms. This fact is sufficient to decide not to seek the explanation of different forms of existence of chemical elements in the colloidal state, because the main thing characterizing the forms of existence is the state of atoms.

84   The notion of the *form of existence of chemical elements* was introduced by me (1921) as an empirical generalization. It concerns the special areas of the thermodynamic fields of the atoms' existence, in which their different manifestations can be observed. According to our present ideas, these manifestations can be reduced to various special complexes of atoms, different for each form of their existence. It is obvious that the forms of existence of chemical elements can be quite numerous, and that not all of them can be observed in the thermodynamic fields of our planet.

The atoms of star systems, for example, must obviously be observed in special states that are impossible on Earth. We see that they are ascribed such special states for instance to explain their spectra (ionized atoms according to Meghnad Saha) or to explain the observed enormous masses of certain stars. To explain the latter fact it is necessary to assume the concentration of thousands and tens of thousands of grams of matter in one cubic centimeter of those stars (A. Eddington).[32] These stellar states of atoms are apparently forms of existence that are absent from the Earth's crust. Other forms lacking on Earth may and must occur in the Sun, in the Sun's corona, in the nebulas, comets, and in the Earth's core.

85 We define living matter as special forms of existence of atoms purely by the empirical method, having had as yet no chance to clearly imagine what changes the atoms undergo on entering it. But the complete correspondence of this form of existence of atoms in the Earth's crust to other, indubitably specific forms, makes us think that further investigations will show the changes undergone by atomic systems as they enter living matter.

Different forms of existence of atoms in the Earth's crust are defined empirically. They are characterized at the same time by: 1.) a special thermodynamic field typical of each form 2.) special atomic manifestations 3.) distinctly different histories of elements,

---

32. For instance, for the star Sirius B, the specific gravity of matter must amount to 53,000. One may think that, according to the dynamic notions of Bohr-Rutherford (these models are known to be only an approximation to reality), the electron orbits will lie closer to the kernel than as it is known for common atoms (F. Tirring, 1925). The observed displacement of the red part of the spectrum of Sirius B confirms its enormous density: the displacements of spectral lines, calculated for bodies of such density on the basis of the relativity theory, correspond to those observed (W. Adams, 1925).

and 4.) a definite relation, often inherent only in the given form, between the atoms of different elements (their paragenesis).

86   *In the Earth's crust* it is possible to pick out *four different forms of existence* through which chemical elements pass in the course of time, and which determine their history. These four forms are the following: 1.) *Rocks and minerals*, with the prevalence of stable and immobile *molecules and crystals* which are combinations of elements 2.) *Magmas* – viscous mixtures of gas and liquids in the state of a mobile mixture of dissociate atomic systems containing *neither molecules nor crystals* known to our chemistry[33] 3.) *Dispersions* of elements, in which separate elements are *free*; not connected with each other. It is quite possible that in this state some of the elements are *ionized*, lacking some of their *electrons*.[34] This is a special state of atoms corresponding to the radiant matter of M. Faraday and W. V. Crookes. And finally: 4.) *Living matter*, in which the state of atoms is obscure; usually we imagine these atoms in the state of molecules, dissociated ion systems or dispersed existence. Such ideas seem to me empirically insufficient. It is quite probable that besides isotopes (§83), the symmetry of atoms (symmetry of atomic fields) also plays a part in living organisms that we ignore.

---

33. One of the forms of magma and maybe an independent form of existence of elements is presented by glass of high temperature and high pressure.

34. These two states of an element may present two different forms of elements.

87 *The forms of existence of atoms* (elements) play the same role in the non-homogeneous equilibria as the independent variables of temperature, pressure, chemical composition, and physical states of matter (phases). Like those mentioned above, the forms of existence of atoms characterize the concentric envelopes of the Earth's crust, which change with depth. Thus we must add to the thermodynamic phase and chemical envelope mentioned above (§81) the special envelopes relating to the form of existence of chemical elements. They may be called *paragenetic* envelopes, as they determine in general the paragenesis of the elements, i.e., the laws of their existence together. The biosphere is one of such paragenetic envelopes – the most well-known and accessible one.

88 The idea that the Earth's crust is composed of certain thermodynamic, chemical, phasic, and paragenetic envelopes is a typical *empirical generalization.* At present it cannot be explained; that is, it is not related to any theory of the Earth's origin or to any models of our ideas about the world. Nevertheless, in view of everything said above, such a composition seems a result of the interaction of cosmic forces on the one hand, and the matter and energy of our planet on the other hand. Besides, the character of the matter – for instance, the quantitative correlation of elements – is not an accidental phenomenon either, and it is not only connected with geologic causes. This empirical generalization will be taken as the basis for our further investigation.

Like any empirical generalization, it should be considered as the first approximation to presenting reality, which is subject to further modifications and additions. As to a great part of the first thermodynamic envelope (and the corresponding envelopes related to other independent variables), as well as the fifth thermodynamic one and lower, our knowledge is based on a very small number of facts and connected with conjectures and extrapolations that break the empirical generalization. That is why our knowledge in this field is quite unreliable and changes quickly in the course of scientific development. As a result of the recent progress in physics, we can expect great new achievements and changes of the prevailing ideas in this field.

The exact border between the envelopes cannot be shown in most cases. Everything shows that the surfaces dividing the envelopes change in the course of time and that sometimes these changes take place quickly. Their form is very complicated and unstable. But for the issues touched upon in these essays, the character of our knowledge does not matter much, because the *biosphere* lies entirely outside these envelopes of the Earth's crust, and our knowledge is based upon an enormous amount of factual material which is free from hypotheses, guesses, conjectures, and extrapolations.

89   Of all the facts determining chemical equilibria, temperature, pressure, and the corresponding thermodynamic envelopes are of special significance. This is because they always exist for all the forms of matter, for all its states and chemical combinations. Our model of the cosmos is always thermodynamic. That is why in the history of the Earth's crust it is important to distinguish the origin of matter and the related phenomena issuing from different thermodynamic envelopes.

From here on I will refer to bodies related to the second thermody-namic envelope (the surface one) as *vadose* phenomena, to those related to the third and the fourth ones (the metamorphic ones) as *phreatic,* and to those related to the fifth one as *juvenile.* Matter from the first and the sixth thermodynamic envelopes does not enter the biosphere or has not been noticed in it.

90 the living matter of the first and the second order in the biosphere

The limits of the biosphere are determined first of all by the *domain of life.* Life can manifest itself only in a certain environment, under certain physical and chemical conditions. This is the environment corresponding to the biosphere. But one can hardly doubt that the field of stability of life exceeds the limits of this environment. We do not even know how far it can go beyond them, as we cannot give a quantitative estimation of organisms' adaptability over the course of geological time. We know that adaptability depends on the course of time, is a function of time, and manifests itself in the biosphere in close relation to the hundreds of millions years of its existence. We do not have these millions of years at our disposal and cannot replace them in our experiments.

All our experiments on living beings are conducted on bodies that in the course of infinite time[35] have adjusted themselves to the environment, the biosphere, and have developed the matter and structure necessary to live in it. We know this matter changes

---

35. "Infinity" is an anthropocentric notion. In reality there surely exist some regularities that have not been perceived yet, a certain duration of the evolution of living matter in the biosphere (more than 2×10^9 years?).

over the course of geologic time, and the limits of these changes are unknown to us and cannot at present be deduced by studying their chemical character.[36] Our principal conclusion is that life in the Earth's crust occupies a *smaller* field of envelopes than the domain of its possible existence, though the investigation of nature has firmly consolidated and is constantly confirming our belief that life *has adjusted itself* to these conditions, and that over the course of centuries organisms have developed various forms of organization that help them exist in the biosphere.

The best way to express this belief, this unconscious empirical generalization underlying all our scientific work, is to assert that *life, by means of gradual, slow adaptation, has captured the biosphere, and that this process has not yet stopped* (§112, §122). The pressure of life (§27, §51) manifests itself by expanding the limits of the domain of life in the field of the biosphere. *The field of stability of life is thus the result of adaptability in the course of time.* It is neither unchangeable nor immobile – its limits give no idea of the possible limits of life's manifestation. Studies in paleontology and ecology show that life is slowly and gradually expanding.

91   The field of existence of living organisms is determined not only by their physical and chemical properties, the character and properties of the surroundings or the organism's adaptability to these

---

36. Very often the limits of life are looked for in the physical and chemical properties of the chemical compounds comprising a living organism (for instance, of the proteins), which coagulate at 60-70 °C. But, in these cases the complexity of the possible adaptability of the organisms is not taken into account. Besides, some proteins in a dry state remain unchangeable at 100 °C (M. Chevreul).

surroundings; it depends greatly on the conditions of *respiration and nutrition*; i.e., on the organisms' active choice of the elements necessary for their life. We have already seen the great significance of organisms' gas exchange, or *respiration*, in establishing their energy regime and the general gas regime of the planet, its biosphere. Together with the *nutrition* of organisms, that is, with the movement of liquid and solid matter from the environment into the autonomous field of the organisms by force of their energy (§82), this gas exchange also determines the regions of their existence. I have already touched upon this phenomenon as I pointed to green organisms catching solar energy (§42). Here we should focus upon it in more detail.

In the phenomena of respiration and nutrition, the main element is the *source* from which organisms draw the matter necessary for their life. From this point of view, organisms are divided into two distinct groups, namely, *living matter of the first order* (i.e., autotrophic organisms whose nutrition is independent of other organisms), and *living matter of the second order* (i.e., heterotrophic and mixotrophic organisms.) The division of organisms into three groups according to their nutrition was introduced in the 1880's by the German physiologist W. Pfeffer. This division is a vast empirical generalization, fraught with various consequences, and its meaning in understanding nature is greater than is usually thought.

Autotrophic organisms make up their bodies entirely from matter of "inert," "dead" nature; all their "organic" compounds containing nitrogen, oxygen, carbon, and hydrogen, which comprise the bulk of their mass, are taken from the mineral kingdom. Heterotrophic organisms feed on organic compounds created by other living beings. Ultimately, the preliminary work of autotrophs is necessary for their existence. For instance, their carbon and nitrogen are to a great extent (or entirely) received from living matter. The food of mixotrophic organisms (with respect to carbon and nitrogen) consists of compounds made up both by living matter and by chemical reactions of dead matter.

92   The question concerning the source from which organisms receive the elements they need for life is more complicated than it seems to be at first sight, but the division pioneered by W. Pfeffer is obviously the crucial feature of living nature as a whole. There is not a single organism in the world that is not connected, at least partially, by its respiration and nutrition, to inert matter. Autotrophic organisms stand out due to the fact that they are independent of living matter concerning *all* chemical elements, and can obtain them from their inert, non-living surroundings, i.e., they take the elements they need for life from certain molecules, from the compounds of elements.

But ultimately, an enormous amount of the biosphere's molecules without which life cannot exist are themselves the product of the biosphere. Without these life would not have existed in the inert environment. This concerns for instance the entire amount of free oxygen – O_2 – and almost all the gases – such as CO_2, NH_3, H_2S etc. The participation of life in creating *natural water solutions* is considerable. But the processes of respiration and nutrition are inseparably connected with these solutions. This natural water (and not chemically pure water) is as necessary for life as is gas exchange.

Taking into account this deep influence of life on the character of the chemical bodies of inert matter in the surroundings of which it manifests itself, we must set limits to the independence of autotrophic organisms. We must withdraw from a very common logical conclusion that the autotroph organisms observed at present could exist on our planet by themselves. Not only do they always originate from autotrophs of the same kind, but they also receive the elements they need for life from such forms of inert matter that would not have existed if life had not created them earlier.

93 So the green autotrophs require the presence of free oxygen to enable their existence. That free oxygen is created by them from water and carbon dioxide. In the inert matter of the biosphere it is always a biochemical product. But moreover, we cannot assert that it is their only required element that is completely dependent upon life for its existence. At present, for instance, W. B. Bottomley has posed the question whether the existence of green water plants necessarily requires complex organic compounds – auxonomes as he calls them[37] – dissolved in water. Although this assertion cannot be considered to be proven completely, it seems quite probable as we gradually get to understand the significance of unnoticed and usually ignored admixtures of organic compounds that are found in any natural water, either fresh or salty. All these organic compounds that exist and spring up in the biosphere each second have a mass of many quadrillion tons or

---

37. These now are called auxins. *[Comment by the editor of the 3rd edition]* Auxins were discovered and named by Frits W. Went in 1926 in the Netherlands. Their discovery revolutionized our understanding of how plants function. *–Ed.*

maybe more. They are created by life, and we cannot state that they owe their origin solely to autotrophic organisms.

On the contrary, everywhere we see that those compounds rich in nitrogen that are created by heterotrophic and mixotrophic organisms are exceedingly important both for nutrition of organisms and for creation of minerals (bitumens). In nature we constantly see the manifestation of these bodies, even without making any chemical analysis. They induce the *foam* of marine and other natural water, they show up as thin iridescent films covering hundreds of thousands, or even millions of square kilometers of water surfaces, and they color marshes, rivers and lakes of the tundra, and black and brown rivers of tropical and subtropical regions. Not a single organism is free from them: not only the aquatic organisms, but also those of the terrestrial green cover, which is incessantly receiving these bodies through rain and dew, and mainly through soil solutions.

In natural waters, the quantity of organic, dissolved (and partly dispersed) compounds ranges widely between 10^{-6} and 10^{-2} %. Their average is apparently close to their percentage in seawater; that is, about 10^{18} to 10^{20} tons. Most likely it exceeds the mass of living matter. Ideas of their importance are slowly penetrating the scientific mind. With old naturalists, we find the understanding of this grand phenomenon sometimes in a rather unexpected context. In the 1870's, the great naturalist J. R. von Mayer pointed out their significance in the composition of medicinal waters and in the general economy of nature. The study of the origin of vadose and phreatic minerals shows that their role is more profound and significant than was shown by von Mayer.

94   But the biochemical genesis of compounds of *inert nature* necessary for the existence of autotrophic organisms does not change their great difference from heterotrophic and mixotrophic organisms. *We shall call autotrophic all organisms that take all the chemical elements they require for life in the present-day biosphere from the inert matter surrounding them, and that do not require ready organic compounds of another organism in order to make up their bodies.*

As usual, while defining a natural phenomenon, we cannot cover the whole phenomenon by a short definition. There are mixtures or dubious cases, as for instance *saprophytes*, which feed on dead and decomposed organisms. But the food of saprophytes consists almost always, if not always, of living microscopic beings that permeate dead bodies and remains of organisms. Accepting the notion of an "autotroph" as limited to the *present-day* biosphere, we render it impossible to be used to draw conclusions about the past of the Earth, about the possible beginning of life on Earth in the form of certain autotrophs. For there is no doubt that for all existing autotrophs (§93), the presence of life's products in the biosphere is vital.

95   The difference between living matter of the first and the second orders is most distinctly expressed by their existence in the biosphere. The field of existence of living matter of the second order, dependent for its existence and its food upon autotrophs, is always wider than the habitation area of the latter. We can distinguish two different groups among the autotrophs: on the one hand, green chlorophyll organisms and on the other hand, the realm of minute, quickly propagating bacteria.

We have already seen that green chlorophyll organisms are the principal mechanism of the biosphere catching solar rays and creating chemical bodies by means of photosynthesis. The energy of these bodies becomes the source of active chemical energy of the biosphere, and to a great extent, of the whole Earth's crust. *The field of existence of these green autotrophic organisms is first of all determined by the field of penetration of solar rays* (§23).

Their mass is enormous when compared to the mass of the rest of living matter – it perhaps comprises about half of it (§46). We look at them as mechanisms capable of catching light radiations of low intensity and making complete use of them. The quantities of green matter created by them have possibly differed throughout time, although this widespread opinion has not yet been proved. The great mass of matter captured by them, their ubiquity, their penetration into every place accessible to solar rays, is often why they are considered to be the main foundation of life. It is assumed that over geologic time, many organisms that make up living matter of the second order have originated from them. And now the existence of the whole animal world, and of a great quantity of plant organisms containing no chlorophyll (such as fungi and bacteria) is entirely conditioned by them.

Their chemical work in the Earth's crust is most important – they create free oxygen by decomposing stable oxygen compounds such as water and carbon dioxide in the process of photosynthesis. They have most probably performed the same work throughout all the remote geologic periods. The phenomena of weathering clearly show us the exclusive role of free oxygen in the Archeozoic era – the same role it plays in the present-day biosphere. As far as we can now judge, the composition of the products of weathering in the Archeozoic era – their quantitative correlations – were the same as at present. Obviously, the source of free oxygen was the same too; the world of green plants. The entire mass of free oxygen was on the same order as we see nowadays. The amount of green matter, and the energy of solar rays generating it at that faraway and alien time of hundreds of millions years ago, could hardly differ from the present state (§57).

We have no remains left of green organisms from the Archeozoic era. They turn up continuously beginning with the Paleozoic era and manifest an unusually fast development of their numerous forms up to the present time. The number of their forms at present must be no less than two hundred thousand species, and the number of all species that have ever existed on our planet – a number which is not accidental – cannot be estimated now because a relatively small number of their fossil species – a few thousand – shows only the incompleteness of our knowledge. It grows rapidly each decade, even each year.

96 Much smaller quantities of living matter are organized in the form of autotrophic bacteria. The existence of green autotrophic organisms became clear at the end of the 18th century, at the beginning of the 19th century, and in the 1840's. Thanks to the work of J. B. Boussingault, J. B. Dumas, and F. V. J. Liebig, it became part of the scientific mind. As for the existence of autotrophic bacteria devoid of chlorophyll and independent of solar rays, the discovery was made at the end of the 19th century by S. N. Vinogradsky and has not yet affected the scientific mind the way it could be expected to.

These organisms play an enormous role in the geochemical history of sulfur, iron, nitrogen and carbon, but they are not very diverse. No more than one hundred species are known, and they cannot be compared to the green plants by their mass or by their importance. Nevertheless they are dispersed everywhere – we find them in soils, in the silt of water reservoirs and in seawater, but nowhere can we find them in quantities comparable to the quantities of autotrophic greenery of the land or the green plankton of the world ocean. Their geochemical energy is much higher than the same energy of the green plants; exceeding the latter by tens and hundreds of times, it is the maximum for living matter. Although the kinetic energy calculated per hectare will, in the long run, be the same both for unicellular green algae and for bacteria, the algae can reach the maximum stationary state in decades, while bacteria at favorable conditions reach it much faster – in 36-48 hours.

97   We have very few observations of the propagation of autotrophic bacteria. According to J. Reinke they propagate slower than do other bacteria – observations of ferric bacteria by N. G. Cholodny do not contradict this assertion. These bacteria divide one or two times a day (Δ = 1 to 2), while for ordinary bacteria such a rate of division may be observed only if the conditions under which they live are unfavorable. For instance, *Bacillus ramosus*, living in rivers and giving no fewer than forty-eight generations a day under favorable conditions, can produce only four generations at low temperature (Marshall Ward, 1925).

Even if such a slowdown of the propagation of autotrophic bacteria in comparison to other bacteria turns out to be a common phenomenon for all of them, their propagation rate will correspond to the *maximum*, but not average speed of transmission of life of green unicellular plants. That is why we could expect their quantities to be much greater than the masses of green organisms, and that the phenomenon we observe in the sea for unicellular algae (§51); that is, their domination over green metaphytes, will exist for bacteria as compared to green protists.

98 But it is not like that. The reason for the small concentration of living matter in this form of life is quite analogous to the reason for the domination of green metaphytes over green protists on land (§49).

For instance, their unusual ubiquity throughout the depths of the ocean and far beyond the layers accessible to solar rays makes one think that the reason for their relatively small quantities in the biosphere, which is manifested for such different varieties as nitrogen, sulfur, or ferric bacteria, must be a reason not of a particular, but of a *general character*. This can be seen in the quite unusual conditions of their *nutrition* (in the conditions that enable their existence).

All of them receive the required energy by oxidizing non-oxidized or partly oxidized compounds of nitrogen, sulfur, manganese, and carbon into their highest degrees of oxidation. But the necessary initial compounds (materials) that are poor in oxygen – the vadose minerals of these elements – can never be gathered in the biosphere in sufficient quantity. The biosphere is on the whole a chemical realm of oxidation; because of the amount of free oxygen – a creation of green organisms. In this environment rich in oxygen, the most stable forms are the most oxidized compounds rich in oxygen.

In this respect, autotrophic organisms must seek surroundings for their existence, and this accounts for the peculiarities of their organization. They can, and some of them like the nitrogen bacteria probably always do, find the energy needed for their life by means of oxidizing lower degrees of oxidation into higher ones, but the number of chemical elements that allow reactions of this kind is limited. Besides, in the presence of excess free oxygen, the same compounds rich in oxygen come into existence without the influence of bacteria, because in such an environment they present stable forms of molecular structures.

99 *Autotrophic bacteria are in a state of a continuous lack of food* – in a state of hunger. This accounts for the numerous adjustments of their life. Everywhere – in muds, in springs, in seawater and in moist soils – we see a kind of secondary equilibrium between bacteria reducing sulfates and autotrophic organisms oxidizing sulfides. The way these secondary equilibria occur everywhere shows the regularity of the phenomenon. Living matter has developed these structures thanks to the enormous pressure of life from autotrophic bacteria (§27) that fail to find a sufficient quantity of ready compounds poor in oxygen in the biosphere. In this case, living matter creates them by itself in the inert environment. In oceans, the same kind of equilibrium is observed between autotrophic bacteria oxidizing nitrogen, and heterotrophic organisms reducing nitrates. It is one of the most magnificent equilibria in the chemistry of the biosphere.

The ubiquity of these organisms is a manifestation of their enormous geochemical energy; of their great speed of transmission of life. The absence of large conglomerations of autotrophic bacteria anywhere is explained by the lack of compounds poor in oxygen in the biosphere where free oxygen is constantly released by green plants. They do not produce great masses of living matter only because it is physically impossible due to the lack of necessary compounds in the biosphere. Between the quantity of matter formed by autotrophic green organisms and that formed by autotrophic bacteria, certain correlations must exist, which are conditioned by the greater importance of geochemical energy of organisms that create free oxygen and prevail in mass.

100 More than once opinions have been advanced that these original, very specific organisms are representatives of more ancient organisms which appeared earlier than green plants. Not long ago, such ideas were put forward by one of the outstanding naturalists and philosophers of our time, the American H. F. Osborn (1918). Observation of their role in the biosphere, however, contradicts this statement. The close connection of these organisms with the presence of free oxygen shows that they depend on the green organisms and solar radiant energy no less than animals and plants containing no chlorophyll and feeding on matter prepared by green plants. But in the biosphere all the free oxygen – the food of these bodies – is the product of green plants.

The same, secondary meaning of these organisms as compared with green plants is shown also by the character of their functions in the economy of living nature. Their importance is great in the biogeochemical history of both sulfur and nitrogen – the two elements that are so necessary for creating the main matter of protoplasm – protein molecules. If the activities of these autotrophic organisms stopped, life would perhaps have been reduced in quantity but it would have remained a powerful mechanism of the biosphere, since the vadose compounds [nitrates, sulfates and gaseous forms of transferring nitrogen and sulfur in the biosphere (i.e., ammonia and hydrogen sulfide)] are constantly created in great quantities independently of life.

Without prejudging the problem of autotrophic nature (§94) and the beginning of life on Earth, we can say that the dependence of autotrophic organisms on green organisms, and their secondary formation in relation to them, is at least quite probable. Everything shows that in these autotrophic organisms we observe forms of life that maximumly increase the utilization of solar rays. We observe an improvement of the mechanism "solar rays-green organism," and not a form of terrestrial life independent from cosmic radiations. A similar manifestation of the same process is the entire heterotrophic realm of animals and fungi, which is infinite in its forms and species.

101 This is clearly seen also in the characteristics of the distribution of living matter in the biosphere, in the domain of life. *This distribution is entirely determined by the field of stability of green plants*, in other words, by the area of the planet that is permeated by solar light. The main mass of living matter is concentrated in that part of the planet embraced by solar light. The more intense the illumination is, the denser is the mass of green plants. Heterotrophic organisms and autotrophic bacteria are also concentrated here, because they are closely connected in their existence with either the products of the life of green plants (first of all free oxygen), or with the complex organic compounds created by them.

From this area illuminated by the Sun, the heterotrophic organisms and autotrophic bacteria penetrate into areas devoid of solar rays and green life. Many of them live only in these dark areas of the biosphere. It is possible to think that they gradually adjusted themselves to the new conditions of life, because morphological study of the fauna of terrestrial caves and ocean depths often shows indisputably that this fauna originated from ancestors who lived in the illuminated parts of the planet.

From the geochemical point of view, the assemblages, or concentrations of life free from green organisms, are especially important. These are for example the *vital bottom film of the hydrosphere* (§130), the lower parts of the *coastal aggregations of ocean life,* and the vital bottom films of terrestrial water reservoirs (§156). We shall see their enormous significance in the chemical history of the planet. It is possible to demonstrate that their existence is most closely connected, in a direct or indirect way, with the organisms of the green fields of life. A morphological study allows us to stipulate, and in some cases to scientifically assume, the genesis of these organisms by way of a paleontological evolution from organisms of the illuminated parts of the planet. Moreover, their everyday existence is based on the radiant energy of the Sun.

The very existence of these bottom films is largely a result of the remains of the upper parts of the ocean falling to the bottom and having no time to decompose or to be eaten by other organisms on their way. The final source of these organisms' energy should thus be searched for in that part of the planet illuminated by solar light. Free oxygen penetrates into the dark depths of seawater, and we do not know its origin otherwise than biochemically – it being created by the work of green plants. Anaerobic organisms characteristic of the lower parts of the bottom film depend most closely on the aerobic organisms and their remains on which they feed.

Everything indicates that these manifestations of life in the dark regions of the planet are in constant development and that their area of existence is increasing.

Apparently, throughout geologic time there has been a constantly renewed penetration of living matter further and further into the azoic area on both sides of the green cover of the planet, and this process takes place at present as well. So we are living in an age of slow expansion of the domain of life.

102   Maybe this expansion manifests itself also as the biochemical creation of new forms of radiant energy by heterotrophic living matter. In the depths of the sea, the *phosphorescence* of organisms becomes more intense; that is, radiation of luminous waves of the same length as the solar radiation that gives energy to life and which through it effects chemical changes of the planet. We know that this secondary luminous radiation, the phosphorescence of the sea surface which takes place incessantly on our planet and embraces simultaneously hundreds of thousands of square kilometers of its surface, allows the green organisms of plankton to do their chemical work also at a time when the radiant energy of the Sun does not reach them.

Is the phosphorescence of the sea depths another manifestation of the same mechanism? Does intensification of life take place here due to the transportation of the cosmic energy of the Sun kilometers below the surface? We do not know this. But we must not forget the fact that deep-water expeditions have come across green living organisms at depths that are much greater than the domain of penetration of solar rays from above. For instance, "Valdivia" came across the living algae *Halionella* in the Pacific Ocean at a depth of about 2 km.

If living matter proved to be capable of transferring solar radiant energy into new places not only as unstable chemical compounds in a thermodynamic envelope corresponding to the biosphere (§82); i.e., as chemical energy, but also as secondary radiant energy, it would still have only an insignificant effect on the expansion of the principal domain of photosynthesis – no larger than the recreation of luminous energy by Man.

This new, man-made radiant energy is used by living matter, but up until now its part in the total cosmic photosynthesis of the planet is insignificant. Eventually, all green living matter is connected with solar rays and determines the domain of life on Earth. In our further account, we shall distinguish this part of first-order living matter and attribute to it all other manifestations of life.

103 the limits of life

The field of stability of life, as we have learned, exceeds the area of the biosphere determined by its characteristic independent variables that are taken into account while studying the physical and chemical equilibria that can take place in it. The field of stability of life determines the area in which life can achieve its full development. It is probably mobile and has no strict limits.

A characteristic feature of living matter is its changeability, its capacity for adjusting itself to the conditions of its surroundings. Due to this capacity, living organisms can adjust themselves even in a few generations to life under conditions that would have been deadly for the previous generations. At present there is no possibility to experimentally confirm this capacity of changing, because we do not have geologic time at our disposal, and without it no experiment can be done.

Living matter, the sum total of living beings, is distinctly different from inert matter because of its mobile equilibrium exerting pressure on the surroundings. But it is not clear how this pressure depends on the length of time. Such a field of stability of life, which is related to the changeability of the organisms, is also *heterogeneous* (i.e., not *homogeneous*). It is divided into the *gravity field*, or the field of larger organisms, and the *field of molecular forces*, which includes minute organisms such as microbes and ultramicrobes, whose life and movement depend not on gravity but on radiations, both luminous and other.

The area of each of these fields is determined by the changeability of the organisms and by their capacity for accommodation, but both factors have not been studied well enough yet.

So, we shall take into account 1) temperature, 2) pressure, 3) the physical phase of the environment, 4) the chemical characteristics of the environment, and 5) radiant energy. These are the most significant features characterizing the fields of stability of life.

104 In this regard we must distinguish between the conditions that life can stand without stopping its functions, that is, under which it suffers but survives, from the conditions under which a living organism can have progeny (i.e., increase the living mass, the active energy of the planet). Perhaps due to the genetic interconnection of all living matter, these conditions are almost similar for all organisms. But this field is much narrower for the green vegetative cover than for heterotrophic organisms.

Life's limit is determined ultimately by the physical and chemical properties of *the compounds* from which an organism is made, and by their destructibility under certain conditions of the environment. But there is a series of cases that show that *before* the compounds are broken up, the mechanisms that are built by them and that determine the functions of life, break up. Both the compounds and the mechanisms they constitute are changing incessantly throughout geologic time and are adjusting themselves to changes in the surroundings. The maximum field of life can be outlined by the survival limits of select organisms.

105 The highest temperature that can be endured by some heterotrophic beings, such as the spores of fungi especially in the latent form of their existence, is about 140 °C; this varies depending on whether the surroundings of the organism are dry or moist.

The experiments of L. Pasteur on spontaneous generation have indicated that raising the temperature to 120 °C in a moist medium does not kill all the microbial spores. In a dry medium the temperature must be more than 180 °C (E. Duclaux).[38] In the experiments of M. Christen, spores of soil bacteria endured a temperature of 130 °C. for five minutes without perishing, and of 140 °C. for one minute. The spores of a certain bacterium described by E. Zettnow endured exposure to a flow of steam for 48 hours (V. L. Omelianski).

Stability is even greater at low temperatures. Experiments at the Jenner Institute in London have indicated stability (in liquid hydrogen) of bacteria spores for 20 hours at a temperature of -252 °C. A. MacFadyen points out that certain microorganisms kept on living in liquid air for many months at a temperature of -200 °C. According to the experiments of P. Becquerel, the spores of fungi in a vacuum did not lose vitality after 3 days at -253 °C. So it should be considered that the *range of 433 °C is at present the maximum heat field* in which some forms of life can exist for some time without the occurrence of death and decomposition. This interval is much smaller for green vegetation. We have no exact data, but it can hardly exceed 160–150 °C. (from 80 to -60 °C).

106 The range of pressure related to the dynamic field of life is probably very large. The experiments of G. V. Khlopin and G. H. Tammann have shown that certain fungi, bacteria, and yeast can undergo a pressure of about 3000 atmospheres without any

---

38. This impression of the participants of the famous debate of Pasteur and G. Pouchet is to my mind of greater significance for determining the maximum temperature of the heat field of life than experiments with pure cultures. It was based on studying the properties of hay extract, which is closer to the compound medium of life of the Earth's crust than our pure cultures.

noticeable changes of their properties. Yeast remains alive at 8000 atmospheres. On the other hand, thc latent forms of life – seeds or spores – can be preserved for a long time in "airless" space, i.e., at pressures of millesimals of an atmosphere. Apparently there is no difference between the heterotrophic and green organisms (spores, seeds).

107   The enormous significance of waves of a definite length of *radiant energy* for green plants has been noted more than once. It underlies the whole structure of the biosphere. Sooner or later green organisms perish without these radiations. At least some heterotrophic organisms and autotrophic bacteria can live in the darkness. But the character of the radiance of this "darkness" (long infrared waves) has not been studied yet. On the other hand, we know the limits of any form of life in the field of *short waves.*

The medium in which very short ultraviolet waves of less than 300 nanometers spread is inevitably lifeless. The experiments of P. Becquerel have shown that these rays, which consist of waves that rapidly oscillate, kill all forms of life in a very short period of time. The medium they find themselves in, which is interplanetary space, is impassable for any forms of life adjusted to the biosphere, though neither temperature, pressure, nor its chemical character prevent life from existing in it.

Considering this connection of the development of life in the biosphere with solar radiation, the most exact and detailed study of the limits of life should be given as much attention as possible.

108   The range of chemical changes that life can endure is indeed
enormous. The discovery of anaerobic organisms by L. Pasteur
has indicated the existence of life in surroundings devoid of free
oxygen, and has considerably expanded its earlier assumed limits.
The discovery of autotrophic organisms by S. N. Vinogradsky has
shown the possibility of life[39] in the absence of pre-existing organic
compounds – in a purely mineral medium. Spores and grains, latent
forms of life, can apparently exist for any period of time in a medium
devoid of gases and water without any harm being done to them.

At the same time, within the thermodynamic field of life's
existence, its different forms can thrive in a variety of chemical
surroundings. *Bacillus boracicola*, living in the hot boracic water
springs of Tuscany, can live in a saturated solution of boric acid,
and the bacteria easily endures a 10 % solution of sulfuric acid at
normal temperature (G. Bargagli Petrucci, 1914). There is a series
of organisms, mainly certain fungi, that live in strong solutions
of different salts that are deadly for other organisms. There are
also fungi living in *saturated* solutions of vitriol [sulfuric acid],
saltpeter [$NaNO_3$], and potassium niobate. The above-mentioned
Bacillus boracicola endures a 0.3 % solution of mercuric chloride,
while other bacteria and infusoria live even in more concentrated
solutions (A. M. Bezredka, 1925). Yeast, for example, can live in
solutions of sodium hydrofluorate, and the larvae of some flies
survive in a 10 % solution of formalin. Certain bacteria are furthermore known to propagate in an atmosphere of pure oxygen.[40]

---

39. Devoid of chlorophyll. *[Comment of the editor of the 3rd edition]*

40. At present bacteria are known that live in atomic reactors in the
presence of penetrating radiation. *(Comment of the editor of the 3rd
edition)*

The field of these phenomena is relatively poorly studied, but the capacity of life forms to accommodate seems boundless indeed. But all this concerns only heterotrophic organisms. The development of green organisms requires the presence of free oxygen (at least dissolved in water). Strong solutions of salts give these organisms no possibility of development.

109   Although some forms of life in a latent state can live in a dry medium devoid of water, the presence of water in a liquid and gaseous state is a necessary condition for the growth and development of organisms – for their manifestation in the biosphere. The geochemical energy of organisms in the form of their propagation passes from the potential into the free form only in the presence of water in which the gases necessary for breathing are dissolved.

The importance of water for green vegetation is clearly seen and has become a part of the common mentality. The foundation of all life – green life – does not exist without water. But lately it has been possible to better understand the mechanism according to which water functions. It became clear how important it is for life whether the reaction of the water solutions in which the organisms live is acidic or basic, and how important the degree and character of their *ionization* are.

The importance of these phenomena is enormous, because the main mass (according to weight) of living matter of the biosphere is concentrated in natural water, and the conditions of life for all organisms are most closely connected with natural water solutions. All organisms consist of water solutions or water sols.[41]

Protoplasm can be considered as a water sol, in which colloidal condensation and changes take place. Everywhere in the liquids of an organism the phenomena of ionization take place, and with the incessant interaction between the natural water solutions and the liquids of organisms living in them, the correlation of ionization between both media is extremely significant.

Thanks to subtle methods of investigation, we can perform a very exact quantitative observation when the ionization changes, and in this way we have a magnificent means for studying the changes of the principal surroundings where life is concentrated.

Seawater contains about 10^{-9} % of H^+ ions. It is weakly basic, and this small prevalence of negative ions over positive ions H^+ is preserved in general without changes – it constantly restores itself in spite of the countless chemical processes taking place in the sea (pH = 8). This ionization is very favorable for the life of sea organisms, whereas small oscillations have either favorable or unfavorable effects depending upon the organisms. Life can exist only within certain limits of ionization, from 10^{-6} % H^+ to 10^{-10} % H^+. Beyond these limits no life is possible in a water solution.[42]

110 The phase of the surroundings is of great significance for the manifestations of life. Apparently, life can survive in a latent state in surroundings of any phase: liquid, solid, gaseous, or in

---

41. The organisms' mass is 60 to 90% (or even more) composed of water, i.e., the organisms are probably 80-100% composed of water solutions and water sols.

42. At present these limits are increased for bacteria.

"airless" space. At any rate, experiments show that seeds can remain without gas exchange for some time, which means in any phase within the limits of the thermodynamic field of life. But the existence of a living organism with fully developed functions is inevitably connected with the possibility of gas exchange (respiration) and the stability of the colloidal systems of which it consists. That is why organisms can occur only in surroundings in which such exchange is possible: in a liquid, colloidal, or gaseous medium. In solid bodies they can be observed only if the medium is loose and porous, enabling gas exchange. Due to the small size of some organisms, dense solid media can also be a substratum for life. But a liquid medium *devoid of gases* – either a solution or a colloid – cannot be a domain of life. This is another manifestation of the exclusive significance of the gaseous state of matter.

111 the boundaries of life in the biosphere

The previous paragraphs show that the domain of life includes the whole biosphere with its structure, composition, and physical conditions of the medium. Life has adjusted itself to its conditions, and there are no places in the biosphere where it could not manifest itself in one way or another. This is certainly true for the usual, normal conditions of the biosphere, but not for temporary, ephemeral infringements of these conditions such as the craters of volcanoes during eruptions and the hot lavas inaccessible for life. But these are insignificant temporary details in its existence.

The same kind of temporary phenomena are the poisonous exhausts of gases (for instance chlorous or fluoric oxygen) accompanying volcanic processes, or hot volcanic mineral pools devoid of life. Long-lasting phenomena, such as thermal springs with temperatures up to 90 degrees C°, turn out to be colonized by peculiar organisms that have adjusted to their conditions.

It is not clear whether terrestrial brines – i.e., solutions containing more than 5 % salts – are lifeless. The largest conglomeration of such lifeless salty water is situated in the Dead Sea in Palestine. But brine pools, even richer in salts than the Dead Sea, are rich in life. Its absence in the Dead Sea is explained by it being rich in bromine, but this is only a hypothesis – a guess which is not based on experiments. Our notion of the Dead Sea may originate from the incompleteness of our knowledge, as its microfauna – partly bacterial – has not been studied properly. Some acidic (sulfuric) or salty natural waters, the ionization of which is less than 10^{-1} % H^+, must be lifeless (§109), but the reservoirs they make up are insignificant.

112 On the whole, we may assume that the terrestrial envelope in which living matter is observed corresponds completely to the field of life's existence. This envelope is continuous like the atmosphere, and in this way differs from such intermittent envelopes as the hydrosphere. But the terrestrial field of life's stability is not entirely filled with living matter. We observe the slow advance of life into new domains, its capturing of these domains in the course of geologic time.

In the terrestrial field of life's stability we must distinguish first of all the field of temporary penetration of living organisms into where they do not perish immediately and secondly, the domain of their long-lasting existence, which is inevitably connected with propagation. The extreme limits of life in the biosphere must be determined by the existence of conditions insurmountable to all organisms. It is enough even for one condition (an independent variable of the equilibrium) to reach the quantity insurmountable for a living being, be it temperature, chemical composition, ionization of the medium, or the wavelength of the radiation.

Of course, such definitions cannot have an absolute character. The property we call the adaptability of the organism, its capacity for protecting itself from harmful conditions of the environment, is enormous. Its boundaries are unknown to us, especially if we take time into account. Setting up these limits on the basis of the possibilities of survival observed now, we inevitably enter the field of extrapolations, which is always slippery and uncertain.

In particular, a person of reason and skillfully directed resolution can reach, directly or indirectly, fields inaccessible to any other living being. Taking into account the unity of all life that we see everywhere while considering life as a planetary phenomenon, such a property of *Homo sapiens* cannot be thought of as accidental. Its existence makes us look more carefully at the stability of life's boundaries in the biosphere.

113 Such a definition of the boundaries of life, based on the possibility of the organisms' existence in their present-day forms and on the ranges of adaptability, clearly characterizes the biosphere as an *envelope*, as the conditions excluding life manifest themselves along the whole surface of the planet simultaneously. That is why it would be sufficient to state only the upper and the lower limits of the field of life.

The *upper boundary* is conditioned by *radiant energy*, the absence of which makes life impossible. The *lower boundary* is connected with reaching such a high temperature that it limits life categorically.

Within the boundaries set up in this way, life embraces (though not completely) one thermodynamic envelope, three chemical, and three phase ones (§88). The importance of the latter – troposphere, hydrosphere and the upper part of the lithosphere – is most clearly expressed by their phenomena, and these will be the basis of our study.

114   Apparently, no forms of life familiar to us at present can leave the boundaries of the stratosphere, or at least its upper parts. Here another paragenetic envelope begins where there are hardly any chemical molecules in their more complex forms. This is the field of the highest rarefaction of matter, even if we take into account the new calculations of Prof. V. G. Fesenkov (1923–1924) that allow greater quantities of matter than were accepted before. Prof. V. G. Fesenkov assumes that at a height of 150–200 km, the stratosphere contains 1 ton of matter in one cubic kilometer.[43] The existence of atoms of this rarefied matter is not only the consequence of its rarefaction (i.e., the lessening of collisions of gaseous particles and thereby the lengthening of their free trajectories), but is also connected with the powerful effects of ultraviolet and maybe other solar rays (perhaps cosmic radiation as well), which freely reach these extreme boundaries of our planet (§8). We know that ultravio-

---

43. Other calculations give figures of at least 1000 times less: one ton per 100 km^3, a kilogram per 200 km^3.

let rays are chemically highly active. In particular, the rays of waves shorter than 200 nanometers (160–180 nanometers) destroy any kind of life, even the most stable spores in a dry or airless environment. There is no doubt that these rays illuminate the remote upper fields of the planet.

115 These ultraviolet rays do not go lower because they are completely absorbed by *ozone*, of which rather great quantities constantly appear in the stratosphere. Ozone is created from free oxygen and maybe water under the influence of those very deadly ultraviolet radiations of the Sun, which it absorbs.

According to C. Fabry and H. Buisson, the ozone of the stratosphere could make up a layer 5 mm thick if it were concentrated in one place in its pure state. But even as dispersed atoms these quantities are sufficient to detain all the radiation harmful to life.

No matter how much ozone is decomposed, it is always being restored, as the rays shorter than 200 nanometers constantly impact the excess of oxygen atoms in the lower part of the stratosphere. The existence of life is protected by the *ozone screen*, which is 5 mm thick, and which forms a natural upper border of the biosphere.

It is interesting that the free oxygen necessary for the production of ozone is created in the biosphere only in a biochemical way – if life ceases it would vanish. *Life, creating the free oxygen in the Earth's crust, creates ozone and protects the biosphere from the deadly short rays from heavenly bodies.* It is obvious that the novel manifestation of life – civilized humans – can protect themselves in other ways and go beyond the ozone screen without any harmful consequences.

116   The ozone screen determines only the upper border of *possible* life
whereas it actually stops much lower in the atmosphere. Green
autotrophic plants, for example, cannot rise above the green
arboreal and herbaceous cover of the land, and there are no green
cells that develop in the aerial medium. Only the green cells of
plankton can rise lightly from the ocean's waves, but this rise is
accidental. Organisms can get higher than trees either mechanically or thanks to developed mechanisms of flying. Extremely
seldom can green organisms penetrate into the atmosphere for
great distances and long periods. The smallest spores for instance,
those of the conifers or cryptogams, are devoid of or poor in
chlorophyll.

There must be vast masses of green organisms that are blown
by the wind and sometimes rise rather high. The mass of living
matter that penetrates into the atmosphere consists mainly of
living matter of the second order. All the flying organisms belong
to it. The green layer of our planet, where solar radiation begins to
turn into terrestrial chemical energy, is situated on the surface of
the land and in the upper layer of the ocean – it does not rise high
into the atmosphere.

Throughout geological time, however, green plants have expanded
their area of existence into the air. Striving to collect as much solar
radiation as possible, green vegetation has penetrated far into
the lower layers of the troposphere – it has risen dozens, even
hundreds of meters above the surface in the form of high trees
and their associations in forest massifs. These forms of life were
developed by organisms most likely in the Paleozoic era.

117 The smallest bacteria and spores, and the flying forms of animals, penetrate into the atmosphere and exist there for a long time. The relative concentrations of this life, especially of latent forms (such as spores of microscopic organisms), can be observed only in the "dust atmosphere;" that is, in the parts of the air cover accessible to terrestrial dust. The dust atmosphere is connected mainly with the land. This dust atmosphere reaches 5 km according to A. Y. Klossovsky (1910), but according to O. Mengel (1922), large quantities of dust rise no higher than 2.8 km. Furthermore, the dust consists primarily of inert matter.

On mountain summits, air contains very few organisms, but they do exist there. L. Pasteur discovered, by using nutritious liquids, that on average no more than 4–5 pathogenic microbes exist there in 1 cubic meter. A. Fleming discovered not more than 1 pathogenic microbe per 3 liters at a height of 4 km. Apparently, in the upper layers the microflora of the air becomes poorer in bacteria and richer in mold and yeast fungi (V. L. Omelianski).

Microflora undoubtedly go beyond the average limits of the dust atmosphere (5 km), but unfortunately exact observations are very scarce here. It may reach the limits of the troposphere (9–13 km), as the movements of air we observe on Earth; (i.e., winds and air currents) attain this height. But reaching these heights above the Earth's surface is hardly of any importance for the planet's history, because the great majority of these organisms are in a latent state and because the organisms are hardly noticeable within the great mass of inert gas, though rarefied, in which they are dispersed.

118   It is not clear whether animals go beyond the boundaries of the troposphere. Of course, sometimes they get rather high, higher than the highest mountains' summits (which are always within the limits of the troposphere), and thus reach its upper border. According to A. Humboldt's observations, a condor can fly as high as 7 km above the Earth's surface, and he also observed flies on the summit of Chimborazo (5882 m).

These observations of A. Humboldt and some other naturalists have been denied by present-day ornithologists studying the birds' migrations on intermediary stations, but the latest studies of A. F. P. Wollaston (1923), a naturalist from the English expedition to Everest, show that some mountain predators rise to or soar around the summits of the highest mountains – higher than 7 km (7540 m). These must be a few particular species of birds. Far from mountain summits and even in mountain regions, birds hardly reach 5 km. The observations of pilots indicate a height of 3 km for the eagle. Butterflies have been observed at a height of 6.4 km, *spiders* at about 6.7 km, and aphids at about 8.2 km. Among the plants, *Arenaria muscosa* and *Delphinium glaciale* were seen on mountains at heights of 6.2-6.3 km (R. W. Hingston, 1925).

119   It is Man who penetrates the farthest into the stratosphere and brings with him forms of life that follow him quite unconsciously and inevitably: on himself, in himself, or in his products. The domain of human penetration is expanding with the development of aeronautics, and its limits are already going beyond the borders determined by the ozone layer. Sounding balloons, which always contain some representatives of life in their material, go to the greatest heights. On December 17, 1913, such a balloon launched in Pavia reached a height of 37.7 km.

In their creations, humans themselves already rise higher than the highest mountains. These limits were almost reached by G. Tissendier (1875) and J. Glaisher (1868) in balloons: 8.6 km by the first and 8.83 by the second. With the development of airplanes, humans have reached the borders of the stratosphere. The Frenchman M. Callizot and the American J. A. Macready (1925) reached 12–12.1 km, and this height is surely to be surpassed soon. The continuous habitations of humans (i.e., their villages are observed at elevations of 5.1–5.2 km in Peru and Tibet), their railways at an elevation of 4.77 km (Peru) and their ploughed fields at an elevation of 4.65 km.

120 To sum it up, it may be asserted that the life that manifests itself in the biosphere reaches its terrestrial limit – the ozone screen – only in the form of its separate entities. On the whole, not only the stratosphere but also the upper layers of the troposphere, are lifeless. There is not a single organism that lives constantly in the medium of air. And only a thin layer of the atmosphere (usually much less than one hundred meters) can be considered to be filled with life.

One can hardly doubt that the conquest of the aerial medium is a new phenomenon in the geologic history of the planet. It has become possible only with the development of terrestrial organisms; first plants (Precambrian period?), then insects, winged vertebrates (in the Paleozoic period?), and birds (since the Mesozoic period). From the most ancient periods there are indications of mechanical transportation of microflora and spores. But only with the appearance of civilized humanity did living matter begin to make significant progress in conquering the whole atmosphere.

The atmosphere is not an independent domain of life. From the biological standpoint its thin lower layers constitute a part of the boundary layers of hydrosphere and lithosphere, whereas only in the latter do they form concentrations – films – of life (§150). The enormous influence of living matter on the history of the atmosphere is connected not with its direct presence in the gaseous medium, but with its gas exchange; the creation of new gases released into the atmosphere and the absorption of gases from it (§42 and §65). Living matter influences the chemistry of the atmosphere by modifying the thin gaseous layer bordering the Earth or the gases dissolved in natural waters. The grand final result of it – the energy of life embracing the whole gaseous envelope of the planet and the ubiquitous presence of various products of life (mainly free oxygen) – takes place because of the properties of gaseous matter and not of living matter.

121 Theoretically, the lower boundary of life on Earth should be as distinct and clear as the upper one, which is determined by the ozone screen. It must correspond to the highest temperature at which a living organism cannot exist and develop at all, depending on the properties of the compounds of which it consists. The temperature of 100 °C is already such an obstacle. This temperature is reached at a depth of 3–3.5 km below the Earth's surface and in some places even at 2.5 km. *On average we can assume that at a* *depth of 3 km or more from the Earth's surface, living organisms in their present form cannot exist.*[44]

Beneath the surface of the sea the layer of 100 °C lies deeper because the average depth of the ocean is 3.8 km, while the temperature of the sea bottom is close to 0 °C. Apparently in these parts of the Earth's crust the maximum temperature that life can endure will occur on average at 6.5–7 km deep, if the Earth's gradient is the same. The increase in temperature is actually faster here, and the layer that life can endure hardly goes beyond 6 km below the ocean surface.

Of course the temperature limit of 100 °C is a theoretical boundary. On the Earth's surface we come across organisms that can propagate at temperatures over 70–80 °C, but organisms adjusted to life at temperatures over 100 °C do not occur here either. Thus the maximum lower boundary of the biosphere can on average hardly exceed 2.5–2.7 kilometers on land and 5–5.5 kilometers in the ocean regions. This boundary is most likely set by temperature and not by chemical composition, as the absence of free oxygen is no obstacle for the development of life. Free oxygen on land vanishes much earlier – it goes several hundred meters down from the Earth's surface at least. No organisms can live at a depth over 500 meters except anaerobic bacteria.

122 But the high temperature of the deeper layers is a purely theoretical boundary of the biosphere, as other factors influence the expansion of life in a much more powerful way. Besides, as has been mentioned (§101), the regions of the planet that are devoid of

---

44. Later, bacteria were found in deeper boreholes in the Delta of the Mississippi River. *(Comment of the editor of the 3rd edition)*

light are taken over by geologically younger organisms, and this conquest has not yet reached its limit. Here we observe the same phenomenon as has been indicated for the upper boundary: over the course of geologic time, life has slowly been approaching its lower limits, but it is far from them as yet. It is farther from the geoisotherm of 100 °C than from the ozone screen. Apparently, green organisms that require light for their development cannot go beyond the limits of the planetary surface illuminated by the Sun. Only heterotrophic organisms and autotrophic bacteria can penetrate as deep as possible.

Penetration of living matter occurs differently on land and in the ocean. In the ocean diffuse animal life penetrates maximumly depending on the bottom topography. Most obviously, its noticeable representatives penetrate no deeper than 7 km. As deep as 6.035 km, a *Hyphalaster parfaiti*, or an echinoid was found. The swimming deep-sea forms of life can enter the greatest ocean depths,[45] but nothing has been found yet on the bottom deeper than 6.5 kilometers.[46] Diffusing bacteria penetrate the complete layer of water (they have been found at a depth of 5.5 km) and are concentrated in the sea mud. Their presence in sea mud on a large scale is not proven but very probable.

123 Terrestrial life does not penetrate so deeply mainly because free oxygen does not diffuse so far into the Earth's crust. In the

---

45. Sea depths reach almost 10 km. Recently a region of 9.95 km deep has been found near the Kuriles. Before that, the depth of 9.79 km, not far from the Philippines, was considered the greatest.

46. Such findings have occurred later. *[Comment of the editor of the 3rd edition]*

ocean, the free oxygen in solution is inseparably connected with the outside atmosphere. Its percentage in relation to nitrogen is always higher in the ocean than the balance of these gases in the atmosphere. Oxygen reaches the greatest depths of the ocean – about 10 km – and any decrease of its content is continually though tardily replenished from the atmosphere due to dissolution and diffusion.

On land, oxygen vanishes at rather shallow depths; it is absorbed by organisms or by compounds – mainly organic – that quickly oxidize. The analysis of waters from a depth close to 1–2 km usually does not show any free oxygen in their gases. There is a clear distinction between vadose water containing free oxygen and phreatic water devoid of it; the nature of this sharp transition is not yet clear.[47]

Free oxygen usually permeates all of the soil and part of the subsoil. The upper boundary of free oxygen in swampy soil and in swamps is closer to the surface. According to H. Hesselmann, the swampy soils of our latitudes cannot contain any free oxygen beyond 30 cm. In subsoils, free oxygen reaches a depth of several meters, and sometimes 10 meters or even more, if it does not meet obstacles such as hard rock that absorb the free oxygen. Its traces can penetrate the upper parts of these rocks, which are always in contact with the water of their surroundings.

Free caverns and cracks accessible to air reach a vertical depth of several hundred meters in exceptional cases. Mines and boreholes created by human civilization go the deepest at present;

---

47. In the majority of cases the indications of free oxygen result from errors of observation.

they can exceed 2 km in vertical direction, but their importance on the scale of the biosphere is close to zero. Besides,
in the overwhelming majority of cases, such artifacts lie above
the level of the ocean. The greatest lowland – the bottom of Lake
Baikal (rich in life), a real fresh-water sea – is more than a kilometer below the level of the ocean.

Even if we take into account anaerobic life, living matter on land
can nowhere reach such depths as it does in the hydrosphere. And
even those depths are far away from the thermodynamic boundary of the theoretical field of life. Apparently, life in the deep layers
of the continents never reaches the average depth of life in the
hydrosphere (3.8 km.). It should be mentioned that new investigations of the origins of oil and hydrogen sulfide greatly lower the
lower boundary of anaerobic life. The genesis of these phreatic
minerals is most obviously biogenic, and takes place at a temperature that is noticeably higher than that of the Earth's surface. But
even if the organisms (bacteria) found here were thermophilic,
they would live at a temperature of 70 °C, which is still very far
from the geoisotherm of 100 °C.

124 So we see that the mass of living matter is greater in the hydrosphere not only because the latter, thanks to its surface area, is
the dominating part of the domain of life, but also because life is
found there throughout its entire volume. On land, an area which
makes up only 21 % of the planet's surface, the domain of life does
not extend beyond 1.5 km down from the Earth's surface, and on
average forms a layer of a few hundred meters in depth. In this
thin layer where living organisms occur, life descends beneath the
level of the sea only in unique cases.

On the planetary scale, life on land stops at sea level, while in the hydrosphere it embraces a layer extending down to 3.8 km.

125 life in the hydrosphere

The phenomena of life in the hydrosphere, chaotic as they may seem, actually show unchangeable features throughout all geologic history, beginning with the Archeozoic era. We must regard them as constant, always existing, actually immutable features of the mechanism of the Earth's crust as a whole, including the biosphere. During all the geologic periods they remain in certain areas of the hydrosphere, in spite of the eternal changeability of both life and the ocean. These mechanisms of the hydrosphere retain the same characteristics throughout the whole course of geologic time.

An investigation of these mechanisms should be based on the density of life, on selecting areas especially rich in life. In the structure of the ocean we can always point out areas that I will call vital films and aggregations. They may be considered secondary subdivisions of the Earth's envelope presented by the hydrosphere, as they are its continuous concentric areas, or can be so in certain periods of its geologic history. These vital films and aggregations most probably constitute the regions of the most intense transformation of solar rays in the ocean. All the phenomena of marine life should be studied in relation to them and within them if we want to comprehend its manifestations in planetary history. Only on this condition is it possible to discover the geochemical effect of life in the hydrosphere.

Beside the density of life, it is important to discover the properties of vital films and aggregations:

1. *From the point of view of the characteristics of the hydrosphere's green living matter and how it is distributed.* In this way, regions of the hydrosphere are indicated in which most of the planet's oxygen is produced.

2. *From the point of view of how the creation of new living matter of the hydrosphere is distributed in time and space.* That is, the course of propagation in the marine vital films and aggregations. Obviously, this phenomenon can give a quantitative notion of the regular transformation of geochemical energy and of its intensity there.

3. *From the point of view of geochemical processes in the films and aggregations in connection with the history of separate chemical elements in the Earth's surface.* This shows how marine living matter affects the geochemistry of the planet. We shall see that the chemical functions of different films and aggregations are unchanging, definite, and diverse.

126 As has been mentioned already (§55), the whole surface of the ocean is covered by green life (plankton). Here free oxygen is produced which, due to diffusion and convection, permeates the entire mass of ocean water to its deepest parts, its very bottom. On the whole, green autotrophic organisms of the ocean are concentrated in its upper layers, no deeper than 100 m. Deeper than 400 m, there are only heterotrophic organisms and bacteria.

The whole surface of the ocean is the region of *chlorophyll plankton*, and in some places large organisms such as seaweeds and grasses dominate. They exist as two distinctly different, though often inseparable types of life. The *seaweeds and grasses* are especially strong in coastal and shallower sea regions of the oceans *(coastal aggregations)*. But in some areas, seaweeds form floating masses in the open ocean, as for example the famous Sargasso Sea in the Atlantic Ocean, which is more than one hundred thousand km² in size *(sargasso aggregations)*.

The principal mass of green life exists as microscopic unicellular organisms concentrated mainly on the surface of the ocean as plankton. This must be a consequence of their great propagation rate. The observed propagation of plankton corresponds to velocity, the value v, which is equal to 250–275 cm/s. (This value can reach thousands of centimeters per second.) For coastal seaweeds this rate is no more than 1.5–2.5 cm/s (it may reach several dozens of centimeters per second). If the occupation of the ocean's surface that corresponds to its radiant energy depended only on the rate v, plankton would occupy an area of the surface of the ocean one hundred times greater than would the large weeds. *The actually observed distribution of these life forms that liberate oxygen is close to the order of this quantity.*

Coastal weeds can occur only in shallower parts of the ocean[48] – in the sea regions. According to Y. M. Shokalsky (1917), the area of "seas"[49] is no more than 8 % of the ocean's surface, but only a very small part is occupied by a large cover of seaweed and sea-grasses. Obviously, 8 % presents an unachievable limit

---

48. If the ocean bottom drops off sharply close to the coast, the layer of seaweeds occupies a very small area.

49. That is, depths more than 1000–1200 m, including shoals.

for coastal seaweed. The role of the floating sargasso aggregations is even less significant. The largest one, the Sargasso Sea, corresponds to 0.2 % of the ocean's surface.

127   Green life, which is seldom observed in the ocean, does not exhaust all the manifestations of life in the hydrosphere. A powerful development of heterotrophic life, quite unusual on land, is very characteristic of the hydrosphere. The general impression given by ocean life is not mistaken – it is animals and not plants that dominate the mass of living matter and mark all the manifestations of living nature concentrated in the ocean. But all this animal life can exist only in the presence of vegetal life. Its distribution is related to the distribution of green vegetal life and to the effects of the presence of the latter. It is the close connection of different representatives of life according to the conditions of respiration and nutrition that causes the concentration of organisms in the ocean – the *vital films and aggregations* so characteristic of it.

128   The percentage of living matter in relation to the whole mass of the ocean is small. In general, seawater may be called lifeless. Even bacteria dispersed in it, both autotrophs (§94) and heterotrophs, constitute insignificant fractions of its weight.

Larger quantities of living organisms are observed only in films and aggregations; these may constitute several percent of the mass of seawater, though only in some places. Usually for the "living" films and aggregations the mass percentage of their content is *more than one*. Such concentrations of life are regions of great chemical activity. Life is in eternal motion, but as the result of its countless changes, stable or almost stable concentrations appear in the hydrosphere; these are places of *stable equilibria*. They are as constant and as typical of the ocean as ocean currents.

Taking into consideration only the most general features of life's distribution in the ocean, we can list no more than *four stable concentrations of life: two films* – the plankton and the bottom film, and *two aggregations* – the coastal and the sargasso ones.

129 The principal, most typical form of life's concentration *is the thin upper living film of plankton rich in green life*. It actually covers the whole ocean surface. Green vegetal life sometimes prevails within the plankton, but the effects of heterotrophic animal organisms on the chemistry of the planet may be as significant. Phytoplankton is always unicellular, but in zooplankton the Metazoa play a very important role. The Metazoa often dominate to an extent we can never observe on land.

So, in oceanic plankton, ova and milt of *fish*, the crustacea, worms, starfish, etc., are often observed in quantities exceeding other kinds of living matter. Usually, according to J. Hjort, the average quantity of organisms of microscopic phytoplankton in a cubic centimeter ranges between 3–15. For microplankton as a whole, this quantity (the maximum one) rises to hundreds of microscopic organisms per cubic centimeter (E. R. Allen, 1919). The number of cells of phytoplankton is usually smaller than the number of entities of animal (heterotrophic) organisms. These numbers include neither bacteria nor nanoplankton.

Eventually, it must be admitted that for the plankton film, the number of microscopic organisms, or independent centers of transferring geochemical energy, should be counted in the hundreds or even thousands per cubic centimeter. The mass of this dispersed living matter is no less than 10^{-3}-10^{-4} % of the whole mass of the ocean (probably it is much more than that). The thickness of this layer, which is mainly located at a depth of 20–50 m., does not exceed several dozens of meters. Sometimes the plankton rises toward the water surface or sinks down. The number of organisms both above and especially below this thin film decreases remarkably. Deeper than 400 m, plankton are extremely dispersed.

So in the ocean, living organisms constitute a thin film that on average equals $n \times 10^{-2}$ of the thickness of the whole hydrosphere. *In the chemistry of the ocean, this part may be considered active and the rest of the water biochemically passive.* The plankton film is obviously an important part of the mechanism of the biosphere in spite of its thinness, as important as the ozone screen with its insignificant percentage of ozone. Its area covers hundreds of millions of square kilometers, and its mass must be on the order of 10^{15}–10^{16} tons.

130 Another aggregation – *the vital bottom film* – can be observed in the sea mud and in the bottom layer of water, which adjoins and permeates it. In its size and volume this thin layer is similar to the plankton film, but it must exceed the latter in mass. The bottom film is divided into two parts. The upper part – the pelogen[50] – is

---

50. I use the term accepted by limnologists; it was introduced by M. M. Solovyov.

located in the region of *free oxygen*. The Metazoa play a significant role in the rich realm of animal life that develops here. We can also observe very complicated relations between the organisms of a biocoenosis,[51] the quantitative aspects of which are just beginning to be studied.

In some places this fauna reaches enormous size. The benthos aggregations of Metazoa per hectare are on the same order as the aggregations of Metaphyta [plants] on land at their best harvests (§58). These muds filled with life and the benthos related to them undoubtedly constitute notable aggregations of living matter as deep as 5 km and maybe deeper. Only beyond 7 km are there indications of the disappearance of animal benthos, and from 4-6 km on there is a distinct decrease in the number of individuals.

Below the benthos there is a layer of bottom mud, which constitutes the lower part of the bottom film. The prevalence of Protozoa is overwhelming there. The dominating role is played by bacteria with their tremendous geochemical energy. Only its thin upper part of several centimeters contains free oxygen; lower down there is a thick layer of mud filled with anaerobic bacteria and burrowed through by numerous and varied animals.

---

51. When I began working with Russian biologists in the late 1980s, I noticed that they (their translators) often used the word *biocoenosis* (also spelled *biocenosis*). The term is in our dictionaries but was unknown to me although I had been a "closet ecologist" for many years – even taught ecology courses. The *Random House Unabridged Dictionary, Second Edition* gives the following definition: "*biocenosis*: a self-sufficient community of naturally occurring organisms occupying and interacting within a specific biotope." Ecologists in America, at least, have long since replaced this term with *community*. It is an important ecological concept. –*Ed.*

All the chemical reactions here take place in a reducing medium. The significance of this thin layer with respect to the chemistry of the biosphere is immense (§141). The thickness of the bottom film including the layer of mud hardly exceeds 100 m, although it may be thicker in the deep regions of the ocean where organisms such as sea lilies develop. Unfortunately, the concentration of life present can currently be estimated only approximately at an average of 10–60 m.

131   Plankton and the bottom film are found throughout the entire hydrosphere. While the surface of the plankton may on the whole be close to the surface of the ocean (i.e., equal to 3×10^8 km²), the surface of the bottom film must be considerably larger, as it follows all the complexities and deformations of the ocean's bottom relief. In some places two other aggregations are added to these films. Their existence is closely connected with the planet's surface rich in free oxygen, and they are filled with green life and inseparable from plankton; these are *the vital coastal and sargasso aggregations.*

The coastal aggregations sometimes include the entire layer of water up from the bottom film, as they are adapted to shallower regions of the hydrosphere as well. Their area on the whole cannot constitute much more than a tenth of the ocean. Their thickness reaches hundreds of meters and in some places it may constitute 500 meters or even a kilometer. Sometimes they form one layer together with the plankton and the bottom film.

The coastal deposits of life always exist in shallower parts of the ocean, such as seas and coastal regions. They depend on the penetration of light and heat radiation of the Sun, on the erosion of continents, on water solutions rich in organic material taken there by rivers, and on the dust blown off the land. It consists of forests of seaweed and sea-grass, of mollusks, coral reefs, lime algae, etc. The total quantity of life here must inevitably be smaller than that in the plankton or bottom films, because life in ocean water of less than one km depth hardly exceeds (if at all) a tenth of the ocean.

132 A special place is in all probability occupied by the sargasso aggregations of life, which are often ignored or explained in different ways. They differ from the plankton aggregations with respect to the character of the fauna and flora, and from the coastal ones through their independence of the erosion of continents and terrestrial living creatures brought down by rivers. Unlike the coastal aggregations, the sargasso ones are oceanic aggregations and are observed on the surface of the deep areas of the ocean, without any connection with the benthos of the bottom film. They have long been considered as secondary formations, made up from the separated parts of coastal deposits of life brought in by winds and ocean currents. The constant, unchanging locations of their existence in the ocean seemed the result of the distribution of winds, currents and still waters.

This view can be found in the present scientific literature, but it is in sharp contradiction to the facts; at least those concerning the most well-studied and largest Sargasso Sea of the Atlantic Ocean. There we find particular fauna and flora indicating that the origin of some of its representatives is the benthos of coastal regions. Most probably, L. Germain (1924) is right when he describes these as coastal living matter having evolved in the course of geological time due to the slow sinking of a disappeared continent or archipelago which had been at the location of the Sargasso Sea.

The future will show whether this explanation can be applied to all the many other aggregations of life of this kind. But the fact remains that there is a type of vital aggregation rich in large vegetal organisms and filled with specific animal forms that is different from the plankton and bottom films, as well as from the coastal deposits. They have not been counted exactly, but apparently the area of the ocean covered by these aggregations is much smaller than the area of coastal deposits.

133   This shows that hardly 2 % of the total mass of the ocean is occupied by aggregations of life. The rest of its mass contains *life in a dispersed state*. The influence of all these aggregations and films is noticeable throughout the entire ocean, particularly in its chemical processes and its gaseous regime. But the organisms that are situated in the intermediate layers of the ocean bring about no essential changes, not even of the quantitative characteristics of the phenomenon. That is why, speaking of the significance of life in the biosphere, we may put aside the main mass of ocean water and take into account only the four domains of concentrated aggregations: the plankton and bottom films, and the coastal and sargasso aggregations.

134   In all these biocoenoses *propagation* takes place in a certain discontinuous rhythm. The rhythm of propagation corresponds to the rhythm of the geochemical work of living matter. The propagation rhythm of films and aggregations determines the changes of their geochemical work for the whole planet.

As was mentioned above, a most characteristic form of living matter of both oceanic films is the prevalence of protists in their mass – the minute organisms with maximum rate of propagation. There have rarely been times when their rate of life transportation – the value v – could have been less than 1000 cm/s at favorable, normal conditions in the intermediate layers. That is why they possess the most intense gas exchange in proportion to their surface area and manifest the maximum kinetic energy per hectare. This means that they are able to yield a maximum concentration of living matter per hectare in a given period of time, and soon reach the limits of their fertility.

Apparently, these quickly propagating protists are of different composition in the plankton and the bottom films. In the bottom films bacteria prevail; they fill the huge masses of remains of larger organisms that become concentrated there. In the plankton film bacteria are shifted to second place, and the first place is occupied by green protists and Protozoa.

135 The Protozoa of the plankton do not constitute the majority of planktonic animal life. Among the animals Metazoa prevail: crustacea, larvae, eggs, and small fish, etc. The propagation rate of the Metazoa is always slower than that of the Protozoa. In some cases their speed of life transmission is measured in fractions of a centimeter per second. For the ocean fish and crustacea of the plankton, the value v apparently does not go lower than several dozens of cm/s. A great quantity of Metazoa, often as large forms, is characteristic of the structure of the bottom film. Their propagation is sometimes even slower than that of the small organisms of plankton. Possibly, organisms with a very low propagation rate can be found here.

Metazoa and Metaphyta characterize the sargasso and coastal aggregations. Protists of all kinds occupy second place and are insignificant in the rate of geochemical processes in these biocoenoses. In these regions, especially in coastal aggregations, Metazoa begin to prevail with increased depth, and eventually they become the principal manifestation of life. Their possible significance is expressed in the presence of coral forests, hydroids, crinoids, and Bryozoa.

136 The regularity of the rhythm of the propagation process is far from being understood by our scientific thought. We know only that propagation is discontinuous and that in the world surrounding us there is a very definite alternation of this phenomenon, which depends on astronomical parameters. This cycle depends on solar illumination, solar heating, the quantity of life, and the character of the environment. The increase of propagation of certain organisms is related to the increase in movement of the atoms that are most necessary for their life. The decrease of propagation brings forth an opposite process. At present the picture of this phenomenon is most clear for us in the plankton film.

137 Changes in propagation in the plankton film are always rhythmical. They correspond to the oscillations of the surroundings of life that recur yearly – the rhythmical movements of the ocean. These movements of the ocean – the alluvial movements, the changes in temperature, salinity, intensity of evaporation, and illumination – all have cosmic origins.

At a certain time during the spring months, the whole ocean is seized by a wave of organic matter production in the form of new organisms. This wave manifests itself in the offspring of almost all higher animals and is reflected in the composition of the plankton. According to D. Johnston (1911), the mass of planktonic plants and animals living in a unit of ocean water reaches its yearly maximum and then decreases again inevitably as the spring equinox and the rise in temperature come. The picture drawn by Johnston concerns our latitudes, but it is essentially correct for the whole ocean and only the forms of its manifestation may be different.[52]

Plankton is a biocoenosis. All its organisms are closely connected with each other in their existence. Often the crustacea (Copepoda) that feed on diatoms dominate, and sometimes the diatoms do, as for instance in the northern part of the Atlantic Ocean. A regular rhythm is observed annually in the northeastern seas of Europe, which are well studied. In the period from February to June (for most fish, in March and April) the plankton film is filled with fish spawn. In spring, after March,

---

52. Vernadsky fails to mention the important effects of *day length* (and *night length*) on reproductive cycles in a great number of plant and animal species. The phenomenon of *photoperiodism* was described in 1920 in a well-known paper by the U.S. Department of Agriculture plant physiologists, W. W. Garner and H. A. Allard. –*Ed.*

the film swarms with siliceous diatoms such as *Biddulphia* or *Coscinodiscus* and later some species of dinoflagellates. By the summer the quantity of diatoms and pyridinea decreases, and they are replaced by Copepoda and other representatives of zooplankton. In September and October there is a new bloom of phytoplankton – diatoms and pyridinea – but this time it is not as intensive. December and especially January are characterized by a decrease of life and a slow-down of propagation.

At our latitudes, from February to June, and for most fish in March and April, the plankton film is filled with fish spawn. During spring in the North Sea it swarms with the siliceous diatom *Biddulphia* and *Coscinodiscus*. In summer it is filled with *Rhizosolenia*, and in autumn with other diatoms and piridinea. The first two months of the year – January and February – are characterized by an impoverishment of life and a slow-down of propagation. This rhythm in propagation rate, typical and constant, and different for different organisms, repeats annually with immutable exactness, as do all the phenomena brought forth by cosmic causes.

138   geochemical cycle of vital aggregations and living films of the hydrosphere

Geochemically, the course of propagation is expressed in a regular rhythm of terrestrial chemical processes. Every vital film and every aggregation of life is a domain that creates certain chemical products. No doubt, it is highly characteristic of all life that the chemical elements that have once entered its cycles hardly leave them at all or they remain in them forever. Still a small part of them is always isolated in the form of new vadose minerals – it is this part that we observe that is created by the chemistry of the sea.

A living plankton film is the principal domain of isolation of virgin [free] oxygen created by green plants. The compounds of nitrogen, which are extremely significant in the terrestrial chemistry of this element, are concentrated here. It is the center of creation of organic compounds of the ocean. Several times in a year, calcium precipitates here in the form of carbonates, and silicon in the form of opals. Eventually they sink to the bottom and concentrate in the bottom film. We see the results of this work geologically concentrated in the thick deposits of sedimentary rocks such as the cretaceous rocks with nanoplankton algae and rhizopods or the siliceous deposits with diatoms and radiolaria.

139 The sargasso and partly the coastal aggregations are similar to the living plankton film because of their chemical products. They are typical for the creation of free oxygen, oxygen compounds of nitrogen, oxygen and nitrogen compounds of carbon, and compounds of calcium. Apparently, a concentration of magnesium is often observed in such places. Magnesium is present in the solid parts of organisms, although to a smaller extent than calcium, and in this way passes directly into the composition of vadose minerals.

These aggregations are much less important for the history of silicon than is the plankton film, though silicon's cycle through living matter is very intense here too.

140 In the history of all the chemical elements in the regions of living aggregations, a twofold process is significant: the passage of the chemical elements through living matter, and their deposition, exiting living matter in the form of vadose compounds. On the whole, the isolation of these bodies during a short (for instance a yearly) life cycle is not noticeable, since the quantity of elements leaving living matter in this period is insignificant. It becomes noticeable only over geological periods of time. In this way, masses of inert solid matter are deposited in the Earth's crust, which exceed by many times the mass of living matter existing on the planet at any given moment.

In this respect, there is a crucial difference between the plankton film and the coastal aggregations of life.[53] In the latter, much greater quantities of chemical elements leave the life cycle than is the case for the plankton film. As a consequence, they leave a more significant trace in the Earth's crust. These phenomena are especially intense in the lower layers of coastal deposits near the bottom vital film, and in the parts that border the land or penetrate into it. In this last case, typical features are precipitation of solid organic compounds of carbon and nitrogen and vaporisation of gaseous hydrogen sulfide connected with the sulfur leaving the given area of the Earth's crust: sulfates from the salt lakes and bays that border the sea reservoirs.

141   In the coastal aggregations, there is no sharp border between the chemical reactions of the bottom and the surface of the sea, like in the open ocean where both these living, chemically active films are

---

53. The phenomena in the sargasso aggregations are not exactly known to us at present.

separated from each other by a layer of chemically inert water of several kilometers. In the coastal aggregations, the borders between the films of the hydrosphere get closer and disappear completely in shallow seas and near the coasts. In this case, the activities of all the living aggregations merge together and make up regions of especially intense biochemical work of different types.

The bottom film is always a domain of the intense manifestation of the chemical work of life. The first place is occupied by the concentrations of organisms possessing the greatest geochemical energy – bacteria. At the same time the chemical conditions of the usual environment suffer a sharp change here because, thanks to the existence of large quantities of compounds – mainly life's products – which voraciously absorb free oxygen, and due to the slow replacement of the free oxygen existing within the ocean surface layer, a reductive medium dominates in the bottom film (the sea mud). This is where anaerobic bacteria reign. Its thin layer of several millimeters constitutes a region of intense biochemical oxidizing processes that give birth to nitrates and sulfates. It separates the upper population of the living bottom films, which is similar in their chemical manifestations to those of coastal aggregations, from the reducing medium of the bottom mud, which has no analogs in other areas of the biosphere.

Actually, due to the constant stirring of the mud by burrowing animals, the balance between the oxidizing and reducing media is broken; biochemical and chemical reactions take place in both directions, increasing the creation of unstable compounds rich in free chemical energy. At the same time, a characteristic feature of the bottom aggregation is the continual concentration of decaying remains of dead organisms, which incessantly fall to the bottom

from the plankton, sargasso, and coastal films and from the intermediate layers of seas and ocean. These remains of organisms are
filled with bacteria, mostly anaerobic ones, and they also increase
the chemical-reducing character of these deposits of life.

142   The bottom concentrations of life, because of the character of
their living matter, play quite a unique role in the biosphere and
are of special importance for creating the biosphere's inert matter.
The principal products that are formed here through biochemical
processes in anaerobic conditions are solid compounds or colloidal compounds that become solid in the course of time. In these
regions all conditions act to facilitate their preservation, because
here they do not oxidize (do not "burn down").

At a relatively small depth in the sea mud, not only aerobic but also
anaerobic life stops. As the remains of life and the debris of inert
matter fall from above, the lower parts of the sea mud become
lifeless, and the chemical compounds formed by life have no time
to pass into gaseous products or enter new living matter. The living
layer of mud never exceeds several meters, though it is constantly
growing from the surface and dying away at the bottom.

"The disappearance" of the remains of organisms, their passage
into gases, is always a biochemical process. In the layers devoid of
life, the remains of organisms slowly change, and in the course of
geological time change into solid or colloidal vadose minerals. The
products from these origins surround us everywhere; changed in the
course of time by chemical processes, they constitute the surface of
the planet in the form of sedimentary rocks that are several kilometers thick. They are gradually transformed into metamorphic rocks,

change even more, and on getting into the magma envelope of thc Earth, become part of massive hypabyssal rocks; phreatic and juvenile minerals that in the course of time enter the biosphere again under the influence of the energy that manifests itself in the high temperature of these layers (§77, 78). They supply this region of the planet with free energy transformed by life into chemical energy, which a green organism has once received in the biosphere in the form of cosmic radiations – solar rays.

143 That is why the vital bottom films with the adjacent coastal aggregations of life deserve special attention as we estimate the chemical work of living matter on our planet. They form thick, chemically active regions of the Earth's crust, which act slowly but similarly throughout geological time. The distribution of sea and land on the Earth's surface gives a notion about the movement of these films in time and space. The geochemical importance of the vital bottom films is great, both for their oxidizing upper parts (mostly benthos) and their lower reducing layers. It increases in places (above 400 m, §55) where these films merge with the coastal aggregations of life and beside their usual products also contain free oxygen and the biochemical products connected with the latter and the biochemical work of life.

The oxidizing medium of the bottom film is clearly expressed in the history of many chemical elements aside from oxygen, nitrogen, and carbon. First of all, it completely changes the history of calcium on Earth. It is very interesting that of all metals, calcium dominates in living matter. Calcium constitutes more than 1 % of the total compostion of living matter, and in some organisms

– especially marine ones – its quantity exceeds 10 % and even 20 %. Through the work of living matter, calcium is separated in the biosphere from the sodium, magnesium, potassium, and iron with which it is connected in the inert matter of the Earth's crust and with which it can be compared to the extent in which they are dispersed. Following the vital processes of organisms, calcium is transferred to carbonates, complex phosphates and much more rarely into calcium oxalates. Within organisms, calcium is transferred into the form of carbonates and complex phosphates, and in slightly modified forms it is preserved also in vadose minerals of biochemical origin.

The ocean, especially the regions of bottom and coastal aggregations of life, is the mechanism that creates the calcium upper layers of the planet that are lacking in the juvenile silicate masses of its crust and in the deep phreatic regions. Annually no less than 6×10^{14} g of calcium is deposited in the ocean as carbonates and no less than 10^{18}-10^{19} g of calcium takes part in the continuous cycle of living matter. The ocean comprises a notable part of all the calcium in the Earth's crust (about 7×10^{23} g) and a very considerable share of the calcium in the biosphere. Calcium is concentrated not only by the organisms of the benthos possessing a considerable life transfer rate such as mollusks, crinoids, starfish, algae, corals, and hydroids. It is concentrated by the protists of sea mud, still more by those of plankton – including nanoplankton – and by bacteria possessing the maximum kinetic geochemical energy of living matter.

By depositing the compounds of calcium that constitute whole mountains – areas of millions of cubic meters in volume – solar energy determines the chemistry of the Earth's crust through the vital activities of organisms by creating organic compounds and free oxygen. Calcium is deposited mainly in the form of carbonates, and partially in the form of phosphates. It is taken into the ocean by rivers from the land, where its principal portion has already passed (in another form) through life (§156).

144 Besides calcium, these regions of vital aggregations influence the history of other widespread elements of the Earth's crust such as silicon, aluminum, iron, manganese, magnesium, and phosphorus in the same way. Not all is clear to us in these complicated natural phenomena, but the great significance of this vital film in the geochemical history of those elements is beyond doubt.

In the history of silicon, the influence of the bottom film manifests itself in the formation of deposits of the remains of siliceous organisms from both the plankton and the bottom film; that is, radiolarians, diatoms, and sea sponges. The process results in the formation of the largest known conglomerates of free silica, the volume of which reaches hundreds of thousands of cubic kilometers. This free silica is inert and almost unchangeable in the biosphere, but due to its chemical character as a free oxygen anhydride, it is an intense chemical factor in metamorphic and magmatic envelopes of the Earth; a bearer of free chemical energy.

Another biochemical reaction that possibly takes place here is the decomposition of aluminosilicates of kaolin structure by diatoms and perhaps bacteria, which leads on the one hand to the formation of the above mentioned deposits of free silica, and on the other hand to precipitation of hydrates of aluminum oxide. Apparently, this process takes place not only in the mud, but judging by the experiments of J. Murray and R. Irvine, also in the argillaceous lees of seawater, which in itself is a product of the biochemical processes of weathering of inert matter on land.

145   Probably, these regions and the corresponding biochemical reactions are as important in the history of iron and manganese. The result of those reactions is beyond doubt: the formation in the Earth's crust of the greatest deposits of these elements that are known to us. Among them are the young Tertiary iron ores of Kertch and the Mesozoic ones of Alsace-Lorraine. This has been proved by the latest works of the Russian scientists (B. V. Perfilyev, V. S. Butkevich, B. L. Isachenko, 1926–1927). These bog iron ores and chlorites rich in iron must undoubtedly have been deposited in close connection with the remains of organisms, but the mechanism of the process is not clear to us. Apparently it is a bacterial process, at least partially.

Throughout geological history, beginning with the Archaic era, the same processes are repeated. In the same way, for instance, the greatest and oldest conglomerates of iron in the iron ores of Minnesota have appeared. Of similar character are the numerous manganese ores and their largest deposits in the Caucasus in the Kutais region.[54] There are transitions between the iron and manganese ores, and at present their analogical precipitation takes place in widespread areas of the sea bottom. This precipitation is probably of biochemical, bacterial origin.

54. The present Georgia.

146 Of the same character is the isolation of phosphorus compounds, which are deposited on the sea bottom under conditions that are not yet quite clear to us. Their connection with the phenomena of life – biochemical processes – is obvious, but the mechanism of the process is unknown. The phosphorus of such phosphorite deposits, which are known to have been formed since the Cambrian period, is obviously of organic origin. It always depends on the vital aggregations of the ocean bottom. In some regions of the sea bottom (for instance, near South Africa), phosphorite concretions are forming even now, though on a considerably smaller scale. Part of this phosphorus had already been concentrated in the form of phosphates by living organisms.

However, the phosphorus of the organisms, which is so indispensable for life, usually does not leave the cycle of life. The conditions under which it does are not clear to us. In addition, everything indicates that, together with the phosphorus of skeletons (solid compounds of calcium), the phosphorus of colloidal solutions and the phosphates of the solutions of organisms also turn into concretions. This takes place under special conditions upon the death of organisms rich in skeletons containing phosphorus. These conditions make the usual changes of their bodies impossible and create a favorable medium for the vital activities of specific bacteria.

At any rate, there is no doubt about the biogenic origin of these formations, about their close connection with the bottom film, and about the constant repetition of such phenomena throughout geologic time. In this way, the largest conglomerates of phosphorus are concentrated, like those revealed by the Tertiary phosphorites of Northern Africa or the southeastern states of North America.

147   Our knowledge of the chemical work of living matter of this film is still incomplete. It is clear that its role is significant in the history of magnesium, in the history of barium, and probably of other chemical elements such as vanadium, strontium, and uranium. Here we face a vast field of phenomena that are little known. More mysteries are contained in another part of the bottom film – its lower part devoid of oxygen. This is the field of anaerobic bacterial life and of physical and chemical phenomena related to the organic compounds that permeate it. These compounds were created in a different chemical environment by specific living organisms that are alien to life's usual surroundings rich in oxygen.

Although we do not quite understand the processes taking place here, and must put forward hypotheses as to a series of problems related to them, we cannot disregard them and must take them into account while estimating the role of life in the mechanisms of the Earth's crust. Two empirical generalizations are obvious: 1) The significance of these mud deposits that are rich in the remains of organisms in the history of sulfur, iron, copper, lead, silver, nickel, vanadium, apparently cobalt, and perhaps other rarer metals, and 2) the repetition of this phenomenon in different geologic epochs, which shows its connection with certain physical and geographical conditions upon the death of sea reservoirs and their biological character.

148   As for sulfur, there is no doubt about the direct participation of specific living organisms in its precipitation, such as bacteria that secrete hydrogen sulfide, or decompose sulfates or organic compounds containing sulfur. The hydrogen sulfide secreted in this process enters numerous chemical reactions and produces sulfuric metals. This biochemical isolation of hydrogen sulfide is a phenomenon characteristic of this field and is observed constantly and everywhere within the sea mud, whereas along the edges it quickly oxidizes into sulfates again.

The biochemical character of the precipitation of other metallic compounds is not clear. There is much evidence for the fact that iron, copper, vanadium, and maybe also other metals that are present here and that combine with sulfur, are produced through decomposition of organisms. On the other hand, it is probable that organic matter of the sea mud possesses a capacity to affect metals, precipitating them from weak solutions, whereas the metals themselves may have no direct relation to living matter.

But in both cases this precipitation of metals would not have taken place, had it not been for the remains of life; that is, if the sea mud had not been a product of living matter, at least its organic part. At present we can observe such processes on a large scale in the Black Sea (fall out of sulfuric iron), and also on a small scale in a number of other places. Their wide development in other geologic epochs can be traced in many cases. In the Permian and Triassic periods in the Eurasia region, great quantities of copper were isolated in this way from solutions or from living matter.

149 Everything mentioned above shows that in all the geologic periods the distribution of life in the hydrosphere was the same, as well as the manifestation of it expressed by the chemistry of the planet. The vital films of plankton and the bottom film, and the aggregations of life in the sea (at least the coastal one) existed for all these hundreds of millions of years and were part of one and the same biochemical apparatus. The constant shifting of land and sea also shifted the chemically active regions made up by living matter to the surface of the planet – the vital films and aggregations of the hydrosphere. In this way they changed places, like spots on the face of the planet.

Ancient geologic deposits show no indications of changes in the structure of the hydrosphere or its chemical manifestations. Meanwhile, in the course of this time, however, the morphological appearance of the realm of life has changed beyond recognition. Obviously this change did not influence either the quantity of living matter, or its average general composition. The morphological changes took place within certain limits, which did not disturb the impact of life on the chemical picture of the world.

And this is true despite the fact that the morphological changes were undoubtedly connected with great (on the scale of the organism) disturbances of chemical character, which concerned both individuals and species. New chemical compounds were created, previous compounds disappeared with the disappearance of species, but this had no noticeable influence on the geological effect of life as a planetary phenomenon. On this scale, even such a great chemical change in the history of calcium, phosphorus, and possibly magnesium, as the creation of the Metazoa skeleton passes unnoticed.

It is very likely that in pre-Paleozoic time, organisms were devoid of this skeleton. This hypothesis, which is considered by many to be a stipulated empirical generalization, in fact explains much of the paleontological history of the organic world. For this phenomenon not to influence the geochemical history of phosphorus, calcium, and magnesium, it is necessary to assume that before the creation of the skeleton of Metazoa, the precipitation of similar compounds of these elements took place on the same scale. This isolation by means of the vital activities of protists – including bacteria – takes place even now, but earlier it must have played a more considerable and exclusive role.

If these two different phenomena cause the biogenic migration of the same atoms, the morphological changes, even highly considerable ones, may exert no new influence on the geochemical history of these elements. Everything indicates that such a development in fact took place in the geological history of the Earth.

150 the living matter of the land

The land presents a picture quite different from the hydrosphere. In fact, it is a continuous vital film present in the soil and the fauna and flora that inhabit it. But in this single film filled with life, we should pay special attention to the aquatic aggregations of living matter on the Earth's surface, the water reservoirs that are sharply different from the land both from a geochemical and from a purely biological standpoint; their geological effect is obviously also quite different.

Life covers the land with an almost continuous film; we find its manifestations in glaciers and snow, in deserts and on mountain peaks. One can hardly speak about lifeless areas of the Earth's surface – we can speak only about temporary lifelessness, about the rarefaction or scarcity of life. Life manifests itself everywhere in one form or another. The land areas poor in life such as deserts, glaciers and snowfields, together hardly constitute 10 % of its surface. All the rest of the terrestrial surface is a vital film.

151 The thickness of this film is very small – for continuous forest areas it rises no higher than a few dozens of meters above the Earth's surface, and in fields and steppes it rises only several meters high. The forests in the equatorial regions form a vital film of 40–50 meters. The highest trees of 100 m or more get lost in the general image of the vegetation and cannot be considered separately from that level. Life penetrates into soil and subsoil only to a depth of several meters.[55] Aerobic life stops at a depth of 1–5 meters and anaerobic life goes on average only a few dozens of meters deep.[56] In general, areas of vital film of dozens of meters (the forest regions) alternate with those of several meters (the herbaceous cover).

The activities of civilized mankind have brought changes into the structure of this film that are not observed anywhere in the hydrosphere. These changes are a new phenomenon in the geological history of the planet, the geochemical effect of which has not been estimated yet. One of its principal manifestations is a dramatic decrease of the forest regions (i.e., of the thickest parts of the film).

---

55. Apparently this comment concerns the higher plants and animals. *[Comment of the editor of the 3rd edition]*

56. Later anaerobic bacteria were found in subterranean waters several kilometers deep. *[Comment of the editor of the 3rd edition]*

152 We are part of this film ourselves, and the modifications of its structure and manifestations during the yearly solar cycle is absolutely clear to us. Green plants prevail here, including grasses and trees. Insects, ticks, and perhaps also spiders dominate within the fauna. Living matter of the second order – animals and heterotrophic plants – generally play a secondary part in the amazing variety of ocean life.

Considerable parts of land, such as tropical forests like the African hyle, or the Northern taiga, are always deserts with respect to mammals, birds, and other vertebrates. The Crustaceae, which are not so noticeable to us, constitute a very much dispersed population of these immense vegetal communities. The seasonal strengthening and weakening of life, which has slowly been clarified for plankton, is trivial here. In winter life dies away at our latitudes, and in spring it awakens and develops. The same process goes on everywhere in different forms, more or less pronounced, from the poles to the tropics.

This phenomenon is not only clearly expressed for green surface vegetation and its corresponding fauna for which the seasonal periods of propagation are as characteristic; the same is going on in the soils. Unfortunately, this aspect of the problem is poorly investigated, although the importance of soils in planetary history is much greater than is usually thought.

In general, all films of the hydrosphere and of the land portray strengthening and weakening of propagation; the course of geochemical energy of living matter, the "whirlwinds" of chemical elements caught by it. These periods are regulated by the Sun. The geochemical processes pulsate – regularly slow down and strengthen. We know nothing about the mathematical regularities that obviously exist here.

153 Geochemical phenomena that are related to the vital film of the land are extremely characteristic and make it quite different from the sea films. In the vital film of the land, the processes of the chemical elements leaving a life cycle never lead to such conglomerates of vadose minerals as we observe in marine deposits, where millions of tons of calcium and magnesium carbonates (limestones and dolomitic limestones), silica (opals, etc), hydrates of iron oxides (bog iron-ores), water oxides, manganese (pyrolysis and psilometanes), and complex calcium phosphates (phosphorites) are annually deposited (§143). All these formations are mainly of marine and at any rate aquatic origin.

In the terrestrial part of the biosphere, even a greater portion of chemical elements remains within the cycle of life (§142) than is the case for the hydrosphere. After the death of an organism or the atrophy of its parts, its matter is either immediately captured by new organisms or goes to the atmosphere as gaseous products. These biogenic gases – O_2, CO_2, H_2O, N_2, NH_3... – are immediately caught by living matter again and become part of its gaseous exchange.

This is a quite perfect dynamic equilibrium, which leads to the fact that the immense geochemical work of terrestrial living matter after dozens of years of its existence leaves behind only inconsiderable traces in the solid minerals that form the Earth's crust. The chemical elements of living matter of the land are in constant motion [cycle], as are gases and living organisms.

154 This dynamic equilibrium annually yields a small percentage of the mass of the solid remains of terrestrial living matter in the form of "organic matter" debris, presumably constituting many millions of tons. It consists mainly of carbon, oxygen and hydrogen compounds, and to a lesser extent of phosphorus, sulfur, iron, silicon, etc., that permeate the whole biosphere and in some as yet indeterminable way leave the cycle of life sometimes for millions of years.

These remains of organisms permeate all matter of the biosphere, both living and inert, gather in all vadose minerals, in all surface waters, and are taken down to the sea by rivers and atmospheric fall-outs [precipitation]. Their influence in the course of chemical reactions of the biosphere is immense and analogical to the influence of organic matter dissolved in natural waters mentioned above (§93). The organic remains in the thermodynamic field of the biosphere are full of free chemical energy, and due to their small size they easily disperse in water and thereby form colloidal solutions.

155 On land the organic remains are concentrated in soils that, however, cannot be considered as inert matter. In soils, living matter reaches several dozens of its percentage of mass. It is the field of the highest geochemical energy of living matter, a geochemical laboratory where most significant chemical and biochemical processes take place. In its significance, soil is similar to the mud portion of the bottom film (§141), but unlike it, the soil is mainly an oxidizing medium; instead of being only several millimeters thick like in the bottom mud, life in the soil can be more than a meter deep. Here too, burrowing animals are a powerful factor in its leveling.

The soil is an area of intense weathering in a medium rich in oxygen and carbon dioxide, which is partly created by the living matter that is present in it. But unlike the biochemistry of the terrestrial surface, the chemical creations of the soil do not completely enter the new vital whirlpools of elements, which, using G. Cuvier's metaphor, express the essence of life. They do not turn into gaseous forms of matter but leave the life cycle for a while and manifest themselves in another great phenomenon of the planet (i.e., in the composition of the natural salty water of the ocean). The soil is alive as long as it is moist; its processes proceed either in solutions or in dispersed systems. This accounts for the character of the manifestation of its living matter in planetary chemistry.

156   The water of the land undergoes a constant cycle. This cycle is driven by the energy of the Sun, its thermal rays. The cosmic energy on our planet manifests itself in this way to an extent no less than it does in the geochemical work of life. The activities of water in the mechanism of the Earth's crust are absolutely determining – they are especially noticeable in the biosphere. Not only does water on average comprise much more than two-thirds of living matter's mass (§109), its presence is an indispensable condition for propagation of living organisms, the manifestation of their geochemical energy, and their participation in the mechanisms of the planet. Not only is water in the biosphere inseparable from life – life is also inseparable from water. It is difficult to note where the influence of water ends and influence of the varieties of living matter begins.

Soil is directly caught in the cycle of water, as it is carried by water during downpours and floods. It is constantly and generally leached – surface waters are flowing over and through it. They are constantly dissolving and carrying away its parts rich in organic matter. The composition of the fresh water connected with the soil is directly determined by its chemistry; it is a manifestation of the soil's biochemistry. In this way the soil is a determining factor in the general composition of rivers where all these surface waters eventually gather.

Rivers carry their waters to the sea, and the composition of salt is mainly controlled by them. It is influenced by the oxidizing character of the soil medium, and this is expressed in the final soluble products of its living matter. In rivers, sulfates and carbonates prevail, and sodium is combined with chlorine. Being closely connected with the biochemistry of these elements in the soil, the character of their presence in river water is sharply different from their solid deposits in the lifeless terrestrial envelopes.

157 The circulation of water on land brings about other regular chemical manifestations of its living matter. Life populating the aquatic spaces is clearly different in its effects from life on the surface. Here we can observe phenomena that are largely analogous to the films and aggregations of the hydrosphere. On a smaller scale it is possible to separate the plankton film, the bottom film, and the communities. Chemical reactions take place in a reducing medium next to an oxidizing medium. Here at last chemical elements leave the life cycle to a greater extent, and the formation of solid products, which later become part of the sedimentary rocks of the Earth's surface, increases. Apparently, this process of precipitation of solid products is connected with the phenomena of the reducing medium, with quick disappearance of oxygen and the further limitation of not only the aerobic, but also the anaerobic life of protozoa. Even with such a general resemblance, the geochemical effect of this terrestrial phenomenon is essentially different from that observed in the hydrosphere.

158 This is accounted for by a distinct difference between the aquatic reservoirs of the land and the hydrosphere. The main chemical difference is the fresh character of the principal mass of water, and the physical difference is the shallowness of the reservoirs. The main mass of terrestrial water in the region of the biosphere is concentrated in pools, swamps, and lakes; not in rivers. Due to the shallowness of the reservoirs, they present one fresh water vital aggregation. Only in fresh water seas like Baikal can we observe separate vital films like in the hydrosphere. But these are exceptions.

Due to this character of lakes, their biogeochemical role is different from the ocean water reservoirs, primarily because the products of precipitation in the lakes are different. Although silica, calcium carbonates, and bog iron ores are formed in the bottom films and the corresponding aggregations of the land reservoirs, they are second to the precipitation of carbonic compounds. Only here, noticeable precipitation of stable vadose carbon-hydrogen-nitrogen compounds that are poor in oxygen takes place. These are all kinds of coal and bitumen: stable forms of vadose minerals made up from the organic carbon compounds that leave the biosphere.

In their final modification in metamorphic fields, carbon is isolated in the free form of graphite. The reason that the stable carbon-nitrogen compounds are formed only in fresh water reservoirs is not clear to us, but it has not changed throughout geological time. In the salty water of the sea, no noticeable conglomerates of such carbon compounds are known to us. We cannot say whether it is a consequence of the chemical character of the medium or of the structure of living nature, but in both cases it is connected to the characteristics of life.

The conglomerates of these types of organic matter are sources of great potential energy, which are, using J. R. von Mayer's metaphor, "fossilized solar rays." Their significance is immense for the history of humanity, and not indifferent to nature. A notion of the scale on which this process manifests itself can be achieved by calculating the quantity of coal we know. Such coal deposits may have formed in proximity to the seas. I consider it almost without doubt that the principal sources of liquid hydrocarbon, or petroleum, should be looked for in the same terrestrial fresh water bodies, since dependence on living matter of the biosphere is proved for the main types of petroleum deposits.

159 the interconnection of the hydrosphere and the terrestrial vital films

From what is mentioned above, it is clear that all life is an indivisible unity with regular connections not only between its parts, but also with the inert matter surrounding the biosphere. But our present-day knowledge is not enough to get a clear picture. It is up to the future to explain the underlying numerical correlations. As for us, we can show only the general contours of the phenomenon. The main fact is the existence of the biosphere during all geologic periods, ever since the oldest manifestations in the Archaic era.

This biosphere has presented in its essential cycles one and the same chemical system. We see that throughout geologic time, under the influence of the inexorable flow of solar radiant energy, this chemical apparatus has been working in the biosphere and has been created and supported in its activities by living matter. This apparatus consists of certain concentrations of life that, ever changing, occupy the same places in the terrestrial envelopes corresponding to the biosphere. These concentrations – the vital films and aggregations – form smaller subdivisions of the terrestrial envelopes. Their concentric character is in general preserved, though they never show a continuous unbroken cover of the planet's surface.

They are chemically active regions of the planet: here the various stable systems of dynamic equilibria of terrestrial chemical elements are concentrated. These are the regions where solar energy flowing around the Earth takes the form of terrestrial free chemical energy. This happens to a different extent for different elements. The existence of these regions of the planet is connected on the one hand with the energy it receives from the Sun and on the other hand with the properties of living matter, which is the accumulator and the transformer of this energy into the terrestrial chemical variety. The properties and the location of specific chemical elements play a great role in this process.

160 All these concentrations of life are closely interconnected. One cannot exist without the other. This connection between different living films and aggregations and their unchangeable character is an eternal and characteristic feature of the functions of the Earth's crust, which has become apparent over geologic time. As there has never existed a geologic period devoid of land, there has also been none with just ocean. It has been only in the abstract imagination of our scientists that the planet was seen as a spheroid covered by ocean, as in the form of *Panthalassa* of E. Suess, or in the form of a dry, levelled dead peneplain, as it was drawn long ago by E. Kant and quite recently by P. Lowell.

The ocean and the land have existed together, ever since the remotest geologic epochs. Their existence is connected with the biochemical history of the biosphere and presents a significant part of its mechanism. From this standpoint, the efforts to trace terrestrial organisms as originating from the marine ones seem absurd and fantastic.[57] Aerial life within geologic time is as old as marine life – its forms are developing and changing, but these changes always occur on the Earth's surface, not in the ocean.

Had it not been so, there would have been a revolutionary period, a period of a sudden change of the biosphere that would have been revealed by geological processes. But nothing of the kind exists.[58] Since the Archaic era, the mechanism of the planet and the biosphere is in general the same.

---

57. Here the author is wrong even for his day. *[Comment of the editor of the 3rd edition]*

58. Contemporary geology sees such periods in the processes of metallgeny and types of sedimentation. *[Comment of the editor of the 3rd edition]*

Life remains constant in its main features – only its form undergoes changes. In fact, all the vital films have always existed – the plankton, the bottom, the soil films, as well as all its aggregations – the coastal, the sargasso, and the fresh water communities.

Their interrelations, the quantity of matter engaged in them, had changed, fluctuating in the course of time. However, these fluctuations could hardly be significant, since with the same or almost the same flow of energy of solar radiation throughout geological time, the distribution of this radiation in the films and aggregations had to be determined by living matter – the principal and only changeable part in the thermodynamic field of the biosphere.

But living matter is not an accidental creation – it reflects solar energy in itself. The analysis can be continued, directed deeper into the complicated mechanism as it is presented by all living films and aggregations, and into the chemical interrelations that must reveal themselves between them. I hope to tackle two problems in my following essay – that of homogeneous living beings and that of the structure of living nature in the biosphere.

a few words about the noösphere

1 World War II is coming to its crucial point [1944]. It started anew
in Europe in 1939, after a break of twenty-one years, and has now
been going on for five years in Western Europe and for three years
here in Eastern Europe. In the Far East it was resumed earlier, in
1931, and has been going on for 13 years. In human history and in
the biosphere in general, a war of such intensity, duration, and
force is an unprecedented phenomenon.

As well, World War I (1914–1918), although less intense, was closely
connected with it causally. In this country World War I resulted
in a historically new form of state organization, not only in the
economic sphere, but also in the sphere of national aspirations. A
naturalist (and I think also a historian) should and must consider
historic events of such intensity as a single, extensive, terrestrial,
geological, and not only historical process.

As for my own scientific work, World War I influenced it crucially. It
dramatically changed my geological understanding of the world.
In the atmosphere of that war, I approached a by-then-forgotten
understanding of nature in geology that was new for me and for
others. It concerned a geochemical and biogeochemical approach
embracing inert and living matter from one and the same point
of view.

2 I spent the years of World War I in constant creative scientific
activity and continue in the same direction up till now. In 1915,
now 28 years ago, the academic "Committee for Studying the

Productive Resources" of this country was founded in the Russian Academy of Sciences in Petersburg. The so-called CSPR (of which I was Chairman) played a notable role during the critical time of the world war. It was quite unexpected for the Academy to learn during the war that in tsarist Russia there were no exact data on strategic resources – as they are called now – and we had to rapidly sum up the dispersed pieces of information and bridge the gaps in our knowledge.

With a geochemical and biogeochemical approach to studying geological phenomena, we covered all surrounding nature through one and the same atomic aspect. This was the point (which I did not realize) coinciding with the main characteristic of the science of the 20th century and distinguishing it from the science of the previous centuries. The 20th century is the century of scientific atomism. All these years, no matter where I was, I was captured by the thought of geochemical and biogeochemical manifestations in the surrounding nature – the biosphere. Observing it, I directed my reading about and reflection on the same subject intensely and systematically. I stated my results gradually in lectures and reports in the cities I happened to live in at that time: Yalta, Poltava, Kiev, Simferopol, Novorossiysk, Rostov, etc. In addition, in all these cities I read everything I could get on this subject for its general understanding.

Standing on empirical ground, I put aside all philosophical searches as much as possible and tried to base myself only on exactly stated scientific and empirical facts and generalizations, sometimes allowing scientific working hypotheses. This should be held in view further on. In this connection, I introduced into the phenomena of life the concept of "living matter" instead of that of "life." Now this concept seems to firmly establish itself in science. "Living matter" is an aggregate of living organisms. It is the scientific, empirical generalization of all the known, easily and precisely observed, innumerable empirically indisputable facts.

The concept of "life" always goes beyond the limits of the concept of "living matter" in, for example, the sphere of philosophy, folklore, religion and art. All this is excluded from the concept of "living matter."

3 In the midst of the intense and complex present-day life, Man practically forgets that he himself, and the whole of mankind from which he cannot be separated, are closely related to the biosphere – the particular part of the planet they live on. Their connection with its matter-energy structure is geologically regulated.

Man is commonly referred to as an individual, freely living and moving over our planet and freely building his history. Up till now, historians, and humanitarians in general, and to some extent biologists, consciously disregarded the natural laws of the biosphere, the only terrestrial envelope where life can exist. Naturally Man cannot be separated from it. And this inseparable connection is becoming clear to us only at present.

In reality, not a single living organism exists on the Earth in a free state. All these organisms are constantly and inseparably connected with their matter-energy surroundings by their nutrition and respiration. They cannot exist outside it in natural conditions.

In the year of the Great French Revolution (1789), the outstanding academician from Petersburg who had devoted all his life to Russia – Caspar Friedrich Wolff (1733–1794) – expressed this distinctly in his book *On Special and Active Force Characteristics of Vegetative and Animal Living Substance*, that was published in German in St. Petersburg. He based his approach on Newton, like the overwhelming majority of biologists of his time, and not on Descartes, .

4 Mankind as living matter is inseparably connected with the matter and energy processes of a certain geological envelope of the Earth – its biosphere. Not for a single moment can mankind exist independently of the biosphere. The concept of "biosphere" (i.e., the "realm of life") was introduced into biology in the early 19th century by Lamarck (1744–1829) in Paris, and into geology by E. Suess (1831–1914) in Vienna at the end of the same century. In our century, the biosphere acquires quite a new understanding. It is seen as a planetary phenomenon of cosmic character. In biogeochemistry, we must take into account the fact that life (living organisms) actually exists not only on our planet, not only in the biosphere of the Earth. It has been stated as a possibility, to my mind indisputably, for all the so-called "Earth-like" planets (i.e., for Mars, Earth, and Venus).

5 In the Biogeochemical Laboratory of the Moscow Academy of Sciences, which has recently been renamed Laboratory of Geochemical Problems, together with the academic Institute of Microbiology (headed by Corresponding Member of the Academy of Sciences, B. L. Isachenko), we posed the problem of cosmic life as early as 1940 as a current scientific problem.

In connection with the war, this work was suspended but will be renewed at the first opportunity. In the archives of science, our science included, the idea of life as a cosmic phenomenon has existed for a long time. Centuries ago, in the late 17th century, the Dutch scientist Christiaan Huygens (1629–1695) put forward this problem scientifically in his last book *Kosmotheoros*, which was published after his death. This book was published twice in Russian on the initiative of Peter the Great under the title "The Book of World-Seeing." In the first quarter of the 18th century, Huygens made the scientific generalization that "life is a cosmic phenomenon somehow distinct from inert matter." Recently I called this generalization "the principle of Huygens." Regarding its mass, living matter comprises a tiny part of the planet. Apparently, this observation holds throughout geological eternity.

Living matter is concentrated in the thin, more or less continuous film on the terrestrial surface; in the troposphere; in forests and fields; and it permeates the whole ocean. Its quantity is measured by fractions not larger than tenths of a percent of the biosphere in weight, on the order of about 0.25 %. On land it does not make up continuous conglomerations and goes no deeper than 3 km. It does not exist outside the biosphere. In the course of geological time it undergoes natural morphological changes. The history of living matter in the course of time manifests itself in the slow modification of life forms, forms of living organisms that are continually connected with one another genetically, from generation to generation without any breaks.

For centuries this problem was raised in scientific research, and in 1859 it eventually received a firm substantiation in the great achievements of Charles Darwin (1809–1882) and A. Wallace

(1822–1913). It took shape in the teachings of the evolution of species, plants and animals, including Man. The evolutionary process is inherent only in living matter. The inert matter of our planet does not equally manifest such a process. In the Cryptozoic era, the same minerals and rocks took shape as nowadays. The only exceptions, then and now, are the biologically inert natural compounds (materials) that are always associated with living matter in one way or the other.

The change of the morphological structure of living matter observed in the evolutionary process throughout biological time inevitably leads to change of its chemical composition. This problem now demands experimental checking. We have included this problem into a research proposal for 1944, together with The Paleontological Institute of the Academy of Sciences.

6 If the quantity of living matter is insignificant in comparison to the inert and bio-inert masses of the biosphere, the biogenic rocks (i.e., those created by living matter) comprise an enormous part of its mass and go far beyond the limits of the biosphere. Taking into account the phenomena of metamorphism, they lose all traces of life, become a granite envelope, and leave the biosphere. The granite envelope of the Earth is the realm of bygone biospheres. According to Lamarck's book *Hydrogéologie*, which is noted for many brilliant ideas, living matter, as I understand it, was the creator of the basic rocks of our planet. J. B. Lamarck de Monet (1774–1829) did not accept the discoveries of Lavoisier (1743–1794) up till his death. But another outstanding chemist, his younger contemporary J. B. Dumas (1800–1884), who worked a lot on the chemistry of living matter, had long considered the quantitative significance of living matter in the structure of the rocks of the biosphere.

7 The younger contemporaries of C. Darwin, J. D. Dana (1813–1895) and J. Le Conte (1823–1901) – two prominent North American geologists (Dana is also a mineralogist and a biologist) – had discovered an empirical generalization before 1859 which showed that the evolution of living matter proceeded in a certain direction. This phenomenon was called "cephalization" by Dana and "Psychozoic Era" by Le Conte. Like Darwin, J. D. Dana came to this conclusion, to this understanding of living matter, during his voyage around the world, which he began two years after C. Darwin's return to London in 1838, and which lasted until 1842.

It should be noted here that the expedition to study coral islands, etc., during which Dana came to his conclusions about cephalization, was historically very closely related to the investigations of the Pacific Ocean, including the oceanic travels of Russian sailors such as A. Johann von Krusenstern (1770–1846). Published in German, they encouraged the American John Reynolds (a lawyer) to make efforts to organize a similar American scientific marine expedition. He began to press for it in 1827, when the description of Krusenstern's expedition appeared in German. Thanks to his persistence, the expedition took place eleven years later in 1838. This was the expedition of Charles Wilkes, which eventually proved the existence of Antarctica.

8 Empirical ideas of the direction of the evolutionary process, without any attempts to substantiate them theoretically, date back to the 18th century. Buffon (1707–1788) already spoke about the realm of Man in which he lived, on the basis of the geological significance of Man.

The evolutionary idea was alien to him. It was alien also to L. Agassiz (1807–1873), who introduced the concept of a glacial period into science. Agassiz lived during the time of a rapid rise of geology. He thought that the realm of Man had come geologically, but his theological ideas made him contend against the theory of evolution. Le Conte pointed out that Dana, who had previously shared the views of Agassiz, eventually accepted evolution in the common understanding of the concept, at that time Darwin's concept. The difference between Le Conte's concept of "Psychozoic Era" and Dana's "Cephalization" disappeared.

Unfortunately, this significant empirical generalization has not yet drawn the attention of biologists, especially in this country. The truth of Dana's principle (Le Conte's "Psychozoic Era"), which escaped the view of our paleontologists, can easily be checked by those who want to do this in any contemporary book on paleontology. Not only does it embrace the whole animal realm, but it also manifests itself clearly in specific types of animals.

Dana pointed out that in the course of geologic time (i.e.. throughout at least two billion years and probably much more), an uneven perfection or growth of the central nervous system (brain) is observed. It begins with the Crustaceae – on which Dana empirically stated his principle – and mollusks (squids) and ends with Man. This phenomenon he called Cephalization. The level of brain (central nervous system) attained in the course of evolution never regresses: it only progresses.

9   Proceeding from the geological role of Man, A. P. Pavlov (1854–1929) spoke during the last years of his life of the anthropogenic era in which we are living. He never considered the possibility of the destruction of spiritual and material values we are living through now as a consequence of the barbaric invasion of the Germans and their allies, but he correctly stressed that Man was becoming a growing powerful geological force before our very eyes. This geological force developed over a geologically long time and quite imperceptibly for Man. It coincided with the changes (mainly material) of Man's position on our planet.

In the 20th century, for the first time in history, Man knew and embraced the biosphere, completed the geographical map of the Earth, and settled all over its surface. Mankind has become a single entity in its life. There is not a single corner of the Earth where Man could not survive if necessary. Our stay on the floating ice of the North Pole in 1937–1938 has indisputably proved this. At the same time, thanks to powerful technology and the success of scientific thought, thanks to radio and television, Man can speak to anybody in any part of the planet. Flights and transport have reached the speed of several hundred kilometers per hour, and this is not the limit. All this is a result of Dana's Cephalization (1856), of the growth of the human brain and of the work guided by it.

The economist, L. Brentano illustrated the planetary significance of this phenomenon with an impressive image. He calculated that if every person was given a square meter, and if all the people were arranged side by side, they would not even occupy the whole area of the small Bodensee [Lake Constance] on the border of Bavaria and Switzerland. The rest of the Earth's surface would remain free of people. Thus the whole of mankind comprises an insignificant mass compared to total planetary matter. Mankind's power is connected not with its matter but with its brain, its thoughts and its work guided by its mind. In the geological history of the biosphere, a great future is opened to Man if he realizes it and does not direct his mind and work to self-destruction.

10 The geologic evolutionary process corresponds to the biological unity and equality of all people in *Homo sapiens* and his geological ancestors such as *Sinanthropus*, whose progeny for white, red, yellow, and black races – among all of them – developed incessantly throughout countless generations. It is a natural law: All races interbreed and give fruitful progeny. A historic contest, such as the present war, is eventually won by the party that follows this law. I use the notion of a "natural law" here as an exactly stated empirical generalization as is now common for the physical and chemical sciences.

The historic process is changing dramatically before our eyes. For the first time in history the interests of the human masses – of all and everyone – and the free thought of an individual, determine the life of mankind, become the standard of its notions of justice. Mankind taken as a whole is becoming a powerful geological force. Humanity's mind and work face the problem of reconstructing the biosphere in the interests of freely thinking mankind as a single entity. This new state of the biosphere that we are approaching without noticing it is "the Noösphere."

11   In 1922–1923, in my lectures at the Sorbonne in Paris, I took biogeochemical phenomena as the basis of the biosphere. Some of these
lectures were published in my book "Essays on Geochemistry."
The French mathematician and Bergsonian philosopher, Edouard
Le Roy, proceeded from the biogeochemical foundation of the
biosphere that I had stated. In 1927, during his lectures at the
College de France in Paris, he introduced the concept of "the
Noösphere" as a present-day stage that is geologically developing
within the biosphere. He stressed the fact that he had come to this
notion together with his friend, a prominent geologist and paleon-
tologist, Teilhard de Chardin, who is at present working in China.

12   The noösphere is a new geological phenomenon on our planet. In
the noösphere, Man becomes a great geological force for the first
time. He can and must reconstruct the realm of his life with his
work and mind. More and more creative opportunities are opened
to him. Maybe the generation of my granddaughter will approach
their full blossom. Here we are facing another riddle. If thought is
not a form of energy, how can it modify material processes? This
question has not been solved up till now. As far as I know, the first
to pose it was the American scientist born in Lvov, the mathematician and biophysicist, Alfred Lotka, but he could not solve it.

Goethe (1749–1832), who was not only a great poet but also a great
scientist, was right to say that in science we can know only how
things occur, but not why and what for. In many instances we see
the empirical results of such an "incomprehensible" process. The
mineralogical rarity, pure iron, is now produced in the billions
of tons. Native [pure] aluminum, which has never existed on our

planet, is produced in any desired quantity. The same concerns the numerous newly created artificial chemical compounds on our planet (biogenic cultured minerals). The mass of such artificial minerals is increasing steadily; all the strategic resources belong to it.

The planet's face – the biosphere – is consciously and mainly unconsciously being chemically and physically changed by Man. As the result of the growth of human culture in the 20th century, the coastal seas and parts of the ocean are changing more and more dramatically. Man must take more and more steps now to preserve the riches of the sea for future generations.

Besides, Man is creating new species and races of plants and animals. Like some fantastic dream, we see the future in which Man is striving to go beyond the limits of his planet – to space. And he will probably succeed. At present we cannot disregard the fact that in the great historic tragedy we are experiencing now, we went the right way, which corresponds to the noösphere. Only a historian and a statesman approach the perception of the phenomena from this standpoint. In this respect, the approach of Winston Churchill to this problem (1932) is very interesting.

13   The noösphere, the last of the many states of the evolution of the biosphere throughout geological history, is the state of the present. The course of this problem is just beginning to become clear from studying its geological past in some of its aspects. Here are some examples. Five hundred million years ago, in the Cambrian geological era, the skeleton structures of animals rich in calcium appeared for the first time in the biosphere, and those of plants appeared more than two billion years ago. This calcium function of living matter which has now developed so powerfully, was one of the most important evolutionary stages of geological change of the biosphere.

A change of the biosphere as important as this occurred 70–110 million years ago during the Cretaceous, and especially the Tertiary era. At that time, our green forests, so dear and close to us, were first created in the biosphere. This is another great evolutionary stage analogous to the noösphere. Apparently, in these forests Man appeared by way of evolution about 15 to 20 million years ago.

Now we are going through a new geological evolutionary change of the biosphere. We are entering the noösphere. We are entering this new spontaneous geological process at a terrible time, at the time of a destructive world war. But the important thing for us is the fact that the ideals of our democracy correspond to a spontaneous geological process, to natural laws – to the noösphere. So we can look at the future with confidence. It is in our hands. We shall not let it go.

index

For further information on Vladimir I. Vernadsky and other
biospheric books from Synergetic Press, please visit our website:
www.synergeticpress.com

Synergetic Press

1 Bluebird Court
Santa Fe, New Mexico 87508
Tel. (505) 424-0237
Fax. (505) 424-3336